Anonymous

The Scientific American Reference Book

Vol. 1

Anonymous

The Scientific American Reference Book
Vol. 1

ISBN/EAN: 9783337407339

Printed in Europe, USA, Canada, Australia, Japan

Cover: Foto ©berggeist007 / pixelio.de

More available books at **www.hansebooks.com**

THE
SCIENTIFIC AMERICAN
REFERENCE BOOK

A COMPENDIUM OF

USEFUL INFORMATION

Containing the Complete Census of the United States
Counties; Map of the United States; Views of Public
Buildings; Patent and Copyright Laws; Rules and
Directions for Obtaining Patents; Forms for Patents, Caveats, Assignments, and Licenses;
Hints on the Value and Sale of Patents;
Tables of the Weights and Measures
of the United States; The Principal Mechanical Movements,
with 150 Diagrams; History
and Description of the
Steam - Engine,
Valuable Tables,
Calculations,
Problems,
Etc., Etc.

OFFICIAL CENSUS

OF THE
UNITED STATES, BY COUNTIES, FOR 1890.

ALABAMA.—Area, 50,722 square miles.

Autauga....13,330	Cleburne....13,218	Etowah......21,926	Limestone..21,201	Pike......24,423
Baldwin.....8,941	Coffee......12,170	Fayette.....12,823	Lowndes...31,550	Randolph..17,219
Barbour....34,898	Colbert.....20,189	Franklin...10,681	Macon......18,439	Russell....24,093
Bibb........13,824	Conecuh....14,594	Geneva.....10,690	Madison....38,119	Saint Clair.17,353
Blount.....21,927	Coosa.......15,906	Greene......22,007	Marengo...33,095	Shelby.....20,886
Bullock....27,063	Covington ..7,536	Hale........27,501	Marion......11,347	Sumter.....29,574
Butler.....21,641	Crenshaw...15,425	Henry......28,847	Marshall...18,935	Talladega..29,346
Calhoun....33,835	Cullman...13,439	Jackson.....28,026	Mobile.....51,587	Tallapoosa.25,460
Chambers..26,319	Dale........17,225	Jefferson...88,501	Monroe.....18,990	Tuscaloosa.30,352
Cherokee...20,459	Dallas......49,350	Lamar......14,187	Montg'm'y.56,172	Walker.....16,078
Chilton....14,549	De Kalb...21,106	Lauderdale23,739	Morgan......24,089	Washington 7,935
Choctaw....17,526	Elmore....21,732	Lawrence ..20,725	Perry......29,332	Wilcox.....30,816
Clarke.....22,624	Escambia..8,066	Lee.........28,694	Pickens....22,470	Winston....6,552
Clay........15,765	Total..1,513,017			

ARKANSAS.—Area, 52,198 square miles.

Arkansas..11,432	Craighead..12,025	Howard.....13,789	Miller.......14,714	Randolph..14,485
Ashley.....13,295	Crawford...21,714	Independ'e21,961	Mississippi 11,635	St. Francis 13,543
Baxter......8,527	Crittenden.13,940	Izard........13,038	Monroe.....15,336	Saline......11,311
Benton....27,716	Cross........7,693	Jackson.....15,179	Montgom'y..7,923	Scott......12,635
Boone.....15,816	Dallas........9,206	Jefferson...40,881	Nevada.....14,832	Searcy......9,664
Bradley.....7,972	Desha......10,324	Johnson....16,758	Newton......9,950	Sebastian..33,200
Calhoun.....7,267	Drew.......17,352	Lafayette....7,700	Ouachita..17,033	Sevier......10,618
Carroll.....17,288	Faulkner..18,342	Lawrence..12,984	Perry........6,538	Sharp......10,418
Chicot......11,419	Franklin...19,934	Lee..........18,886	Phillips.....25,341	Stone.......7,043
Clark......20,997	Fulton......10,984	Lincoln.....10,255	Pike.........8,537	Union......14,977
Clay........12,200	Garland....15,328	Little Riv'r 8,903	Poinsett.....4,272	Van Buren..8,567
Cleburne ...7,884	Grant........7,786	Logan......20,774	Polk..........9,283	Washingt'n32,024
Clevel'nd..11,362	Greene....12,908	Lonoke.....19,263	Pope........19,458	White......22,946
Columbia..19,893	Hempstead22,796	Madison....17,402	Prairie.....11,374	Woodruff..14,009
Conway....19,459	Hot Spring 11,603	Marion.....10,390	Pulaski.....47,329	Yell........18,015
Total..1,128,179				

CALIFORNIA.—Area, 188,981 square miles.

Alameda...93,864	Inyo..........3,544	Monterey...18.637	S. Joaquin.28,629	Sonoma....32,721
Alpine........667	Kern.........9,808	Napa........16,411	S.L. Obispo16,072	Stanislaus..10,040
Amador....10,320	Lake.........7,101	Nevada.....17,369	San Mateo..10,087	Sutter......5,469
Butte......17,939	Lassen.......4,239	Orange......13,589	S. Barbara.15,754	Tehama....9,916
Calaveras...8,882	LosAng'l's101,454	Placer......15,101	S. Clara...48,005	Trinity.....3,719
Colusa.....14,640	Marin......13,072	Plumas......4,933	S. Cruz....19,270	Tulare....24,574
Con. Costa.13,515	Mariposa....3,787	Sacr'm'nto.40,339	Shasta....12,133	Tuolumne ..6,082
Del Norte..2,592	Mendocino.17,612	San Benito..6,412	Sierra.......5,051	Ventura...10,071
El Dorado..9,232	Merced.....8,085	S. Bernar'o25,497	Siskiyou..12,163	Yolo......12,684
Fresno.....32,026	Modoc.......4,986	San Diego..34,687	Solano......20,946	Yuba.......9,636
Humboldt..23,469	Mono........2,002	S. Franc'o298,997	Total..1,208,130	

COLORADO.—Area, 104,500 square miles.

Arapahoe.132,135	Delta........2,534	Hinsdale.....862	Mesa........4,260	Rio Blanco 1,200
Archuleta....826	Dolores.....1,498	Huerfano...6,882	Montezuma 1,529	Rio Grande 3,451
Baca........1,479	Douglas.....3,006	Jefferson...8,450	Montrose...3,980	Routt......2,369
Bent........1,313	Eagle........3,725	Kiowa........1,243	Morgan......1,601	Saguache...3,313
Boulder....14,082	Elbert.......1,856	Kit Carson..2,472	Otero........4,192	San Juan..1,572
Chaffee.....6,612	El Paso....21,239	Lake........14,663	Ouray........6,510	San Miguel 2,909
Cheyenne....534	Fremont.....9.156	La Plata.....5,509	Park.........3,548	Sedgwick...1,293
Cl'r Creek..7,184	Garfield.....4,478	Larimer.....9,712	Phillips......2,642	Summit....1,906
Conejos.....7,193	Gilpin........5,867	Las Animas17,208	Pitkin.......8,929	Washington 2,301
Costilla......3,491	Grand........604	Lincoln......689	Powers......1,969	Weld......11,736
Custer......2,970	Gunnison..4,359	Logan.......3,070	Pueblo....31,491	Yuma........2,596
Total..412,198				

CONNECTICUT.—Area, 4,674 square miles.

Fairfield...150,061	Litchfield...53,542	N. Haven.209,058	Tolland......25,081	Windham..45,158
Hartford..147,180	Middlesex.39,524	N. London.76,634	Total......................................746,258	

4 CENSUS OF THE UNITED STATES.

DELAWARE.—Area, 2,120 square miles.

Kent........22,964 N. Castle...97,182 Sussex......38,647 Total......................168,493

FLORIDA.—Area, 59,268 square miles.

Alachua....22,934	De Soto...... 4,944	Jackson....17,544	Manatee.... 2,895	St. John.... 8,712
Baker....... 3,333	Duval......26,800	Jefferson...15,757	Marion....20,796	S. Rosa...... 7,961
Bradford.... 7,516	Escambia...20,188	La Fayette.. 3,686	Monroe....18,786	Sumter...... 5,363
Brevard..... 3,401	Franklin... 3,308	Lake........ 8,034	Nassau..... 8.294	Suwannee..10,524
Calhoun.... 1,681	Gadsden...11,894	Lee.......... 1,414	Orange......12.584	Taylor...... 2,122
Citrus...... 2,394	Hamilton.. 8,507	Leon.......17,752	Osceola 8.133	Volusia..... 8.467
Clay........ 5,154	Hernando.. 2,476	Levy........ 6,586	Pasco 4.249	Wakulla.... 3,117
Columbia...12,877	Hillsboro...14,941	Liberty...... 1,452	Polk 7,905	Walton...... 4.816
Dade....... 861	Holmes..... 4,396	Madison....14,316	Putnam....11,186	Washington 6.426
Total..391,422

GEORGIA.—Area, 58,000 square miles.

Appling 8,676	Colquitt.... 4,794	Gwinnett...19.899	Meriwether20.740	Stewart 15,582
Baker...... 6,144	Columbia...11.281	Habersham11.573	Miller...... 4,275	Sumter...... 2,107
Baldwin....14,608	Coweta.....22,354	Hall........18,047	Milton 6,208	Talbot......13,:58
Banks...... 8,562	Crawford... 9,315	Hancock ..17,149	Mitchell...10,906	Taliaferro .. 7,291
Bartow.....20,616	Dade........ 5,707	Haralson...11.316	Monroe.....19.137	Tattnall....10,253
Berrien 10,694	Dawson 5,612	Harris...... 6,797	Montgom'y. 9.248	Taylor...... 8,666
Bibb........42,370	Decatur....19,949	Hart........10,687	Morgan......16,041	Telfair...... 5.477
Brooks.....13,979	De Kalb...17,189	Heard....... 9,557	Murray 8,461	Terrell.....14.503
Bryan...... 5,520	Dodge......11,452	Henry......16,220	Muscogee..27.761	Thomas....26,154
Bulloch 13.712	Dooly......18,146	Houston....21,613	Newton.....14,310	Towns...... 4,064
Burke......28,501	Dougherty.12,206	Irwin....... 6,316	Oconee...... 7,713	Troup......20,723
Butts.......10,565	Douglas.... 7.794	Jackson....19,176	Oglethorpe.16.951	Twiggs...... 8,195
Calhoun.... 8,438	Early....... 9,792	Jasper 13.879	Paulding....11,948	Union....... 7,749
Camden.... 6,178	Echols...... 3,079	Jefferson...17,213	Pickens..... 8.182	Upson......12,188
Campbell... 9,115	Effingham.. 5.599	Johnson.... 6,129	Pierce...... 6 379	Walker.....13,282
Carroll.....22,301	Elbert......15,376	Jones.......12,709	Pike........16.300	Walton.....17,467
Catoosa.... 5,431	Emanuel ..14.703	Laurens ...13.747	Polk........14.945	Ware........ 8,811
Charlton... 3.335	Fannin..... 8.724	Lee......... 9074	Pulaski.....16.559	Warren 10.957
Chatham...57,740	Fayette.... 8,728	Liberty....12.887	Putnam....14 842	Washingt'n25,237
Chattah'ee. 4.902	Floyd......28,391	Lincoln 6.146	Quitman.... 4,471	Wayne...... 7,485
Chattooga..11.202	Forsyth....11,155	Lowndes...15,102	Rabun 5,646	Webster.... 5,695
Cherokee...15.412	Franklin...14,670	Lumpkin... 6,867	Randolph..15,267	White...... 6,151
Clarke.....15,186	Fulton.....84,655	McDuffie... 8,789	Richmond.45,194	Whitfield..12.916
Clay........ 7,817	Gilmer..... 9,074	McIntosh... 6,470	Rockdale... 6,813	Wilcox...... 7.980
Clayton.... 8.295	Glascock .. 3,720	Macon......13,183	Schley..... 5,443	Wilkes.....18,081
Clinch...... 6.652	Glynn......12,420	Madison ...11,024	Screven 14,424	Wilkinson..10,781
Cobb.......22,286	Gordon.....12.758	Marion...... 7,728	Spalding....13,117	Worth......10,048
Coffee......10,483	Greene......17,061	Total...1,837,353

IDAHO.—Area, 86,294 square miles.

Ada....... 8,360	Boise....... 3,342	Idaho....... 2,955	Logan...... 4,169	Owyhee.... 2.021
Alturas.... 2,629	Cassia..... 3,143	Kootenai .. 4,108	Nes Perce.. 2,847	Shoshone... 5,382
Bear Lake.. 6,057	Custer..... 2,176	Latah...... 9,173	Oneida..... 6,819	Washington 3,836
Bingham...13,575	Elmore..... 1,870	Lemhi...... 1,915	Total....................84,885

ILLINOIS.—Area, 55,405 square miles.

Adams.....61,888	Du Page...27,256	Jo Daviess.25,101	Mason......16.067	Saline......19,342
Alexander.16,563	Edgar......26,787	Johnson....15,013	Massac.....11.313	Sangamon..61,195
Bond.......14 550	Edwards... 9,444	Kane.......60,061	Menard.....13,120	Schuyler...16,013
Boone......12, 03	Effingham.19,358	Kankakee..28.732	Mercer......18.545	Scott.......10,304
Brown.....11,951	Fayette....23,367	Kendall....12.106	Monroe.....12.948	Shelby.....31,191
Bureau.....35.014	Ford.......17,035	Knox.......38,752	Montg'm'y.30.003	Stark...... 9,982
Calhoun.... 7.652	Franklin...17,138	Lake.......24,235	Morgan....32,636	Stephenson31,338
Carroll.....18.320	Fulton.....43,110	La Salle...80.798	Moultrie....14,481	Tazewell...29.556
Cass.......15.963	Gallatin ...14.935	Lawrence...14,693	Ogle........28,710	Union......21,549
Champaign 42,159	Greene.....23,791	Lee........26,187	Peoria......70 378	Vermilion..49.005
Christian...30,531	Grundy....21,024	Livingston.38.455	Perry.......17.529	Wabash....11.866
Clark......21,899	Hamilton...17.800	Logan......25 489	Piatt.......17,062	Warren.....21,281
Clay.......16,772	Hancock ..31,907	McDono'gh27,467	Pike........31,000	Washingt'n19,262
Clinton....17,411	Hardin..... 7,234	McHenry..26,114	Pope....... 14,0 6	Wayne......23,806
Coles......30,093	Henderson. 9,876	McLean....63 036	Pulaski.....11.355	White......26,005
Cook.....1,191,922	Henry......33,338	Macon......38,083	Putnam..... 4.730	Whiteside..30,854
Crawford..17,283	Iroquois...35,167	Macoupin..40 380	Randolph..25,049	Will........62,007
Cumberl'd..15,443	Jackson....27,809	Madison...51 535	Richland...15,019	Williamson22,226
De Kalb...27,066	Jasper.....18.188	Marion.....24,341	Rock Isl'd..41,917	Winnebago39,938
De Witt...17,011	Jefferson...22 590	Marshall...13,653	St. Clair....66,571	Woodford..21,249
Douglas....17,669	Jersey 14,810	Total..3,826,351

CENSUS OF THE UNITED STATES.

INDIANA.—Area, 33,809 square miles.

Adams......20,181	Elkhart.....39,201	Jefferson...24,507	Noble......23,359	Starke....... 7,339
Allen......66,689	Fayette....12,630	Jennings....14,608	Ohio........ 4,955	Steuben....14,478
Barthol'w..23,867	Floyd......29,458	Johnson....19,561	Orange.....14,678	Sullivan....21,877
Benton.....11,903	Fountain...19,558	Knox.......28.044	Owen.......15,040	Switzerl'nd12,514
Blackford..10,461	Franklin...18,306	Kosciusko..28,645	Parke......20,296	Tippecanoe35,078
Boone......26,572	Fulton......16,746	Lagrange...15,615	Perry......18,240	Tipton.....18,157
Brown......10,308	Gibson.....24,920	Lake........23,886	Pike.......18,544	Union....... 7,006
Carroll.....20,021	Grant......31,493	La Porte...34,445	Porter.....18,052	Vanderb'g.59,809
Cass........31,152	Greene.....24,379	Lawrence...19,792	Posey......21,529	Vermillion.13,154
Clark.......30,59	Hamilton...26,123	Madison....30,487	Pulaski....11,233	Vigo........50,195
Clay........30,536	Hancock...17,829	Marion.....141,156	Putnam.....22,335	Wabash.....27,126
Clinton....27,370	Harrison...20,786	Marshall...23,818	Randolph..28,085	Warren.....10,955
Crawford...13,941	Hendricks..21,498	Martin.....13,973	Ripley.....19,350	Warrick....21,161
Daviess....25,227	Henry......23,879	Miami......25,823	Rush.......19,034	Washingt'n18,619
Dearborn...23,364	Howard.....26,186	Monroe.....17,673	St. Joseph.42,457	Wayne......37,628
Decatur....19,277	Huntingt'n 27,644	Montg'm'y.28,025	Scott....... 7,833	Wells.......21,514
De Kalb....24,307	Jackson....24,139	Morgan.....18,643	Shelby.....25,454	White......15,671
Delaware...30,131	Jasper.....11,185	Newton..... 8,800	Spencer....22,060	Whitley....17,768
Dubois.....20,253	Jay........23,478	Total..2,192,404		

IOWA.—Area, 50,914 square miles.

Adair......14,534	Clay........ 9,309	Hancock... 7,621	Madison...15,977	Sac........14,522
Adams.....12,292	Clayton....26,733	Hardin.....19,003	Mahaska..28,805	Scott.......43,164
Allamakee 17,907	Clinton....41,199	Harrison...2,356	Marion.....23,058	Shelby....17,6 1
Appanoose 18,961	Crawford..18,894	Henry......18,895	Marshall..25,842	Sioux......18,370
Audubon...12,412	Dallas.....20,479	Howard....11,182	Mills......14,548	Story......18,127
Benton.....24,178	Davis......15,258	Humboldt.. 9,836	Mitchell...13,299	Tama.......21,651
Black H'k..24,219	Decatur....15,643	Ida........10,705	Monona....14,515	Taylor.....16,384
Boone......23,772	Delaware...17,349	Iowa.......18,270	Monroe....13,066	Union......16,900
Bremer.....14,630	Des Moines35,324	Jackson....22,771	Montgom'y15,848	Van Buren 16,253
Buchanan..18,997	Dickinson.. 4,328	Jasper......24,943	Muscatine.24,504	Wapello....30,426
BuenaVista13,548	Dubuque..49.848	Jefferson...15,184	O'Brien....13,060	Warren.....18,269
Butler......15,463	Em'met..... 4,274	Johnson....23,082	Osceola..... 5,574	Washingt'n18,468
Calhoun...13,107	Fayette....23,141	Jones.......20,233	Page.......21,341	Wayne......15,670
Carroll.....18,828	Floyd......15,424	Keokuk....23,862	Pala Alto... 9,318	Webster....21,582
Cass........19,645	Franklin...12,871	Kossuth....13,120	Plymouth..19,568	Winnebago 7,315
Cedar......18,253	Fremont....16,842	Lee........37,715	Pocahontas 9,553	Winnesh'k22,528
Cerr' Gordo14,864	Greene......15,797	Linn........45,303	Polk........65,410	Woodbury..55,632
Cherokee...15,659	Grundy....13,215	Louisa......11,873	Pottawat'e..47,430	Worth...... 9,247
Chickasaw.15,019	Guthrie....17,380	Lucas......14,563	Poweshiek.18,394	Wright....12,057
Clarke......11,332	Hamilton...15,319	Lyon........ 8,680	Ringgold...13,556	Total...1,911,806

KANSAS.—Area, 78,418 square miles.

Allen.......13,509	Doniphan..13,535	Hodgeman. 2,395	Montg'm'y.23,104	Saline......17,442
Anderson..14,203	Douglas....23,961	Jackson....14,626	Morris.....11,381	Scott....... 1,262
Arapahoe... —	Edwards... 3,600	Jefferson...16,620	Morton...... 724	Sedgwick..43,626
Atchison...26,758	Elk........12,216	Jewell......19,349	Nemaha....19,249	Sequoyah... —
Barber..... 7,973	Ellis....... 7,942	Johnson....17,385	Neosho....18,561	Seward..... 1,503
Barton.....13,172	Ellsworth.. 9,272	Kansas..... —	Ness....... 4,944	Shawnee....49,172
Bourbon...28,575	Finney..... 3,350	Kearny..... 1.571	Norton..... 10,617	Sheridan... 3,733
Brown......20,319	Foote....... —	Kingman...11,823	Osage......25,062	Sherman... 5,261
Buffalo.... —	Ford....... 5,308	Kiowa...... 2,873	Osborne...12,083	Smith......15,613
Butler......24,055	Franklin...20,279	Labette....27,586	Ottawa....12,581	Stafford... 8,520
Chase...... 8,233	Garfield.... 881	Lane........ 2,060	Pawnee..... 5,204	Stanton.... 1,031
Chautauq'a12,297	Geary......10,423	Leavenw'th38,485	Phillips....13,661	Stevens.... 1,418
Cherokee...27,770	Gove....... 2,994	Lincoln..... 9,709	Pottawat'e.17,722	Sumner....30,271
Cheyenne.. 4,401	Graham..... 5,029	Linn........17,215	Pratt....... 8,118	Thomas.... 5,538
Clark...... 2,357	Grant...... 1,308	Logan...... 3,384	Rawlins.... 6,756	Trego...... 2,535
Clay........16,146	Gray....... 2,415	Lyon........23,196	Reno.......27,079	Wabaunsee11,790
Cloud......19,295	Greeley.... 1,264	McPherson21,614	Republic...19,002	Wallace.... 2,468
Coffey.....15,856	Greenwood.16,309	Marion....20,539	Rice........14,451	Washington22,894
Comanche.. 2,549	Hamilton... 2,027	Marshall...23,912	Riley......13,183	Wichita.... 1,827
Cowley....34,478	Harper.....13,266	Meade...... 2.542	Rooks...... 8,018	Wilson.....15,286
Crawford..30,286	Harvey....17,601	Miami......19,614	Rush....... 5,204	Woodson... 9,021
Decatur.... 8,414	Haskell.... 1,077	Mitchell...15,037	Russell.... 7,333	Wyandotte.54,407
Dickinson.22,273		Total... 1,427,096		

KENTUCKY.—Area, 37,680 square miles.

Adair......13,721	Bath.......12,813	Boyle......12,948	Butler......13,956	Carroll..... 9,266
Allen......13,692	Bell........10,312	Bracken....10,308	Caldwell...13,186	Carter......17,204
Anderson..10,610	Boone......12,246	Breathitt.. 8,705	Calloway...14,675	Casey......11,848
Ballard.... 8,390	Bourbon...16,976	Breckin'ge.18,976	Campbell..44,208	Christian..34,118
Barren.....21,490	Boyd......14,033	Bullitt..... 8,291	Carlisle.... 7,612	Clark......15,434

6 CENSUS OF THE UNITED STATES.

Clay........12,447	Greenup....11,911	Lawrence...17,702	Mercer......15,034	Rowan...... 6,129
Clinton...... 7,047	Hancock.... 9,214	Lee............ 6,205	Metcalfe.... 9,871	Russell..... 8,136
Crittenden.13,119	Hardin.....21,304	Leslie........ 3,904	Monroe......10,989	Scott........16,546
Cumberl'd.. 8,452	Harlan...... 6,197	Letcher...... 6,920	Montg'm'y.12,367	Shelby......16,521
Daviess....33,120	Harrison...16,914	Lewis......14,803	Morgan......11,249	Simpson....10,878
Edmonson. 8,005	Hart.........16,439	Lincoln.....15,962	Muhlenb'g.17,955	Spencer..... 6,760
Elliott...... 9,214	Henderson.29,536	Livingston.. 9,474	Nelson......16,417	Taylor...... 9,353
Estill.......10,836	Henry......14,164	Logan........23,812	Nicholas...10,764	Todd.......16,814
Fayette....35,698	Hickman...11,637	Lyon......... 7,628	Ohio.........22,946	Trigg........13,002
Fleming...16,078	Hopkins... 23,505	McCracken21,051	Oldham...... 6,754	Trimble..... 7,140
Floyd......11,256	Jackson...... 8,261	McLean..... 9,887	Owen........17,670	Union.......18,229
Franklin ...21,267	Jefferson ..188,598	Madison....24,348	Owsley...... 5,975	Warren....30,158
Fulton......10,005	Jessamine..11,248	Magoffin 9,196	Pendleton..16,346	Washingt'n13 622
Gallatin.... 4,611	Johnson....11,027	Marion......15,648	Perry........ 6,331	Wayne......12,852
Garrard....11,138	Kenton.....54,161	Marshall...11,287	Pike..........17,378	Webster....17,196
Grant......12,671	Knott....... 5,438	Martin...... 4,209	Powell...... 4,698	Whitley....17,590
Graves.....23,534	Knox........13,762	Mason......20,773	Pulaski.....25,731	Wolfe....... 7,180
Grayson...18,688	La Rue..... 9,433	Meade....... 9,484	Robertson.. 4,684	Woodford..12,380
Green......11,463	Laurel......13,747	Menifee..... 4,666	Rockcastle. 9,841	Total...1,858,635

LOUISIANA.—Area, 41,255 square miles.

Acadia......13,231	Concordia..14,871	Latourche..22,095	Richland...10,230	Tensas......16,647
Ascension.19,545	De Soto....19,800	Lincoln.....14,753	Sabine...... 9,390	Ter. Bonne 20,167
Assumpti'n19,629	E. B. Rouge25,022	Livingston.. 5,769	St. Bernard 4,326	Union.......17,304
Avoyelles..25,112	E. Carroll..12,362	Madison....14,135	St. Charles. 7,737	Vermilion..14,234
Bienville...14,108	E.Felician17,903	Morehouse.16,786	St. Helena. 8,062	Vernon...... 5,903
Bossier.....20,330	Franklin ... 6,900	Natchito's..25,836	St. James ..15,715	Washington 6,700
Caddo......31,555	Grant....... 8,270	Orleans....242,039	S.J. the B'pl1,359	Webster....12,466
Calcasieu..20,176	Iberia......20,997	Ouachita...17,985	St. Landry 40,250	W.B.Rouge 8,363
Caldwell... 5,814	Iberville...21,848	Plaquemi'612,541	St. Martin.14,884	W. Carroll. 3,748
Cameron... 2,828	Jackson..... 7,453	Pt. Coupee.19,613	St. Mary ...22,416	W.Felici'a .15,062
Catahoula..12,002	Jefferson ..13,221	Rapides.....27,042	St. Tamm'y10,160	Winn........ 7,082
Claiborne.. 23,312	Lafayette...15,966	Red River.11,318	Tangipahoa12,655	Total...1,118,587

MAINE.—Area, 31,766 square miles.

Androsc'n..48,968	Hancock...37,312	Lincoln......21,996	Piscataquis16,134	Waldo......27,759
Aroostook..49,589	Kennebec..57,012	Oxford......30,586	Sagadahoc..19,452	Washingt'n44,482
Cumberl'd 90,949	Knox........31,473	Penobscot..72,865	Somerset ...32,627	York........62,829
Franklin...17,053	Total..661,086			

MARYLAND.—Area, 11,124 square miles.

Allegany...41,571	Caroline....13,903	Frederick..49,512	Montgom'y 27,185	Talbot......19,736
A. Arundel34,094	Carroll.....32,376	Garrett.....14,213	P. George ..26,080	Washingt'n39,782
Baltimore ..72,909	Cecil........25,851	Harford....28,993	Qu'nAnne..18,461	Wicomico..19,930
Balti'eCy..434,439	Charles 15,191	Howard.....16,269	St. Mary...15,819	Worcester.19,747
Calvert..... 9,860	Dorchester.24,843	Kent.........17,471	Somerset ..24,155	Total...1,042,390

MASSACHUSETTS.—Area, 7,800 square miles.

Barnstable.29,172	Dukes...... 4,369	Hampden.135,713	Nantucket.. 3.263	Suffolk....484,780
Berkshire..81,108	Essex......209,995	Hampshire51 859	Norfolk....118,950	Worcester 280,787
Bristol....186,465	Franklin ...38,610	Middlesex431,167	Plymouth..92,700	Total...2,238,943

MICHIGAN.—Area, 56,248 square miles.

Alcona..... 5,409	Clare....... 7,558	Iron......... 4,432	Manitou..... 860	Oscoda...... 1,904
Alger...... 1,238	Clinton....26,509	Isabella.... 18,784	Marquette..39,521	Otsego...... 4,272
Allegan....38,961	Crawford .. 2,962	Isle Royal... 135	Mason......16,385	Ottawa 35,358
Alpena.....15,581	Delta......15,330	Jackson....45,031	Mecosta.....19,697	Presque Isle4,687
Antrim....10,413	Eaton......32,094	Kalamazoo 39,273	Menominee33,639	Roscommon 2,033
Arenac.... 5,683	Emmet..... 8,756	Kalkaska... 5,160	Midland...10,657	Saginaw....82,278
Baraga.... 3,036	Genesee....39,430	Kent.......100,922	Missaukee.. 5 048	St. Clair...52,105
Barry......23,783	Gladwin.... 4,208	Keweenaw..2,894	Monroe.....32,337	St. Joseph..25,856
Bay.........56,412	Gogebic....13,166	Lake.........6,505	Montcalm..32,637	Sanilac.....32,589
Benzie..... 5 237	G'd Trave'e13,355	Lapeer 29,213	M'ntmor'cy 1,487	Schoolcraft. 5,818
Berrien....41,285	Gratiot....28,668	Leelanaw.. 7,944	Muskegon..40,013	Shiawassee 30,952
Branch....26 791	Hillsdale...30,660	Lenawee...48 448	Newaygo ...20,476	Tuscola.....32,508
Calhoun...43,501	Houghton..35,389	Livingston.20 858	Oakland....41,245	Van Buren.30,541
Cass........20,953	Huron......28,546	Luce........ 2,455	Oceana......15,698	Washtenaw42,210
Charlevoix. 9,686	Ingham....37,666	Mackinac.. 7,830	Ogemaw.... 5,583	Wayne....257,114
Cheboygan.11,986	Ionia........32,801	Macomb...31,813	Ontonagon. 3,756	Wexford...11,278
Chippewa..12,019	Iosco.......15,224	Manistee ...24,230	Osceola....14,630	Total...2,093,889

MINNESOTA.—Area, 95,274 square miles.

Aitkin..... 2 462	Becker..... 9,401	Benton..... 6 284	Blue Earth 29,210	Carlton..... 5 272
Anoka..... 9,881	Beltrami... 812	Big Stone.. 5,722	Brown......15,817	Carver.....16,532

CENSUS OF THE UNITED STATES. 7.

Cass...... 1.247	Grant...... 6,875	Lyon....... 9,501	Pine...... 4,052	Steele...... 13,232
Chippewa.. 8,555	Hennepin.185,294	McLeod... 17,026	Pipe Stone. 5,132	Stevens...... 5,251
Chisago... 10,359	Houston ...14,653	Marshall ... 9,130	Polk........ 30,192	Swift....... 10,161
Clay........ 11,517	Hubbard... 1,412	Martin...... 9,403	Pope........ 10,032	Todd....... 12,930
Cook........ 98	Isanti....... 7,607	Meeker.... 13,456	Ramsey...139,796	Traverse ... 4,516
Cottonwood 7,412	Itasca....... 743	Mille Lacs.. 2,845	Redwood... 9,386	Wabasha.. 16,972
Crow Wing 8,852	Jackson..... 8,924	Morrison...13,325	Renville... 17,099	Wadena..... 4,053
Dakota.... 20,240	Kanabec.... 1,579	Mower.... 18,019	Rice 23,968	Waseca.... 13,313
Dodge..... 10,864	Kandiyohi..13,097	Murray...... 6,692	Rock........ 6,817	Washingt'n25,992
Douglas.... 14,606	Kittson..... 5,387	Nicollet ... 13,382	St. Louis ..44,862	Watonwan.. 7,746
Faribault..16,708	Lac-qui-p'e.10,382	Nobles...... 7,958	Scott........ 13,831	Wilkin....... 4,346
Fillmore ..26,338	Lake......... 1,299	Norman.... 10,618	Sherburne.. 5,908	Winona..... 33,797
Freeborn...17,962	Le Sueur...19,057	Olmsted....19,434	Sibley...... 15,199	Wright ...24,164
Goodhue...28,806	Lincoln...., 5,691	Otter Tail ..34,232	Stearns 34,844	Y. Medicine 9,854
Total.............				1,301,826

MISSISSIPPI.—Area, 47,156 square miles.

Adams......26,031	De Soto 24,183	Lafayette..20,553	Noxubee....27,338	Sunflower.. 9,384
Alcorn......13,115	Franklin...10,424	Lauderdale29,661	Oktibbeha..17,694	Tallahatc'e 14,361
Amite......18,198	Greene...... 3,906	Lawrence..12,318	Panola......26,977	Tate......... 19,253
Attala......22,213	Grenada ...14,974	Leake........ 20,040	Pearl River 2,957	Tippah..... 12,951
Benton...... 10,585	Hancock ... 8,318	Lee.......... 20,040	Perry....... 6,494	Tishomingo 9,302
Bolivar....29,980	Harrison...12,481	Leflore..... 16,869	Pike........ 21,203	Tunica..... 12,158
Calhoun....14,688	Hinds ,....39,279	Lincoln....17,912	Pontotoc.. 14,940	Union...... 15,606
Carroll......18,773	Holmes 30,970	Lowndes....27,047	Prentiss....13,679	Warren 33,164
Chickasaw.19,891	Issaquena 12,318	Madison ... 27,321	Quitman .. 3,286	Washingt'n40,414
Choctaw....10,847	Itawamba..11,708	Marion....... 9,532	Rankin....17,922	Wayne...... 9,817
Claiborne..14,516	Jackson.....11,251	Marshall...26,043	Scott....... 11,740	Webster....12,060
Clarke......15,826	Jasper..... 14,785	Monroe.....30,730	Sharkey.... 8,382	Wilkinson.17,592
Clay........ 18,607	Jefferson ..18,947	Montg'm'y.14,459	Simpson ..10,138	Winston....12,089
Coahoma ..18,342	Jones....... 8,333	Neshoba .. 11,146	Smith....... 10,635	Yalobusha..16,629
Copiah.....30,233	Kemper....17,061	Newton 16,625	Sumner..... —	Yazoo....... 36,394
Covington.. 8,299	Total...........			1,289,600

MISSOURI.—Area, 67,880 square miles.

Adair......17,417	Clay....... 19,856	Iron.......... 9,119	Montgom'y16,850	St. Clair ...16,747
Andrew....16,000	Clinton....17,138	Jackson ...160,510	Morgan..... 12,311	S.Genev'a.. 9,883
Atchison ..15,533	Cole......... 17,281	Jasper...... 50,500	N. Madrid.. 9,317	S. Francois 17,347
Audrain....22,074	Cooper 22,707	Jefferson ..22,484	Newton....22,108	St. Louis ...36,307
Barry......22,943	Crawford ..11,961	Johnson....28,132	Nodaway..30,914	S.LouisCy451,770
Barton 18,504	Dade........17,526	Knox........ 13,501	Oregon 10,257	Saline......33,762
Bates...... 32,.23	Dallas...... 12,647	Laclede 14,701	Osage 13,080	Schuyler ...11,249
Benton 14,973	Daviess....20,456	Lafayette..30,184	Ozark....... 9,795	Scotland... 12,674
Bollinger...13,121	De Kalb.....14,539	Lawrence..26,228	Pemiscot .. 5,975	Scott....... 11,228
Boone.......26,043	Dent........ 12,149	Lewis 16,935	Perry...... 13,237	Shannon.... 8,718
Buchanan..70,00	Douglas... 14,111	Lincoln...18,346	Pettis....... 31,151	Shelby......15,642
Butler...... 9,964	Dunklin...15,085	Linn........ 24,121	Phelps 12,636	Stoddard ...17,327
Caldwell...15,152	Franklin ..28,056	Livingston.20,668	Pike........ 26,321	Stone........ 7,090
Callaway ...25,131	Gasconade.11,706	McDonald..11,283	Platte...... 16,248	Sullivan ...19,000
Camden....10,040	Gentry.....19,018	Macon......30,575	Polk........ 20,339	Taney...... 7,973
C.Girard'u.22,060	Greene......48,616	Madison... 9,268	Pulaski..... 9,387	Texas....... 19,406
Carroll......25,742	Grundy.. 17,876	Maries 8,600	Putnam....15,365	Vernon 31,505
Carter 5,799	Harrison ..21,033	Marion.....26,233	Ralls....... 12,.94	Warren..... 9,913
Cass.........23,301	Henry......28,235	Mercer.....14,581	Randolph..24.893	Washingt'n13,153
Cedar......15,620	Hickory.... 9,453	Miller......14,162	Ray........ 24,215	Wayne......11,727
Chariton...26,254	Holt........ 15,469	Mississippi10,134	Reynolds.. 6,633	Webster....15,177
Christian..14,017	Howard....17,371	Moniteau..15,630	Ripley...... 8,332	Worth...... 8,738
Clark........15,126	Howell.....18,618	Monroe....20,790	St. Charles.22,977	Wright.....14,484
Total..............				2,679,184

MONTANA.—Area 143,776 square miles.

Beaver H'd 4,655	Dawson ... 2,056	Gallatin.... 6,246	Madison ... 4,692	Park........ 6,881
Cascade.... 8,755	Deer Lodge 15,155	Jefferson.... 6,026	Meagher... 4,749	Silver Bow.23,744
Choteau... 4,741	Fergus..... 3,514	LewisdCl'k19,145	Missoula ..14,427	Yellowstone 2,065
Custer...... 5,308	Total...............			132,159

NEBRASKA.—Area, 75,995 square miles.

Adams......28,303	Brown..... 4,359	Cheyenne.. 5,693	Deuel........ 2,893	Furnas...... 9,840
Antelope...10,399	Buffalo.....22,162	Clay........ 16,310	Dixon 8,084	Gage........36,344
Arthur...... 91	Burt........11,069	Colfax 10,453	Dodge......19,260	Garfield ..., 1,659
Banner..... 2,435	Butler 15,454	Cuming..... 12,265	Douglas...158,008	Gosper...... 4,816
Blackbird.. —	Cass........24,080	Custer...... 21,677	Dundy 4,012	Grant........ 458
Blaine...... 1,146	Cedar....... 7,028	Dakota..... 5,386	Fillmore... 16,022	Greeley ... 4,869
Boone 8,888	Chase....... 4,807	Dawes...... 9,722	Franklin... 7,693	Hall........ 16,513
Box Butte.. 5,494	Cherry...... 6,428	Dawson.... 10,129	Frontier... 8,497	Hamilton..14,096

CENSUS OF THE UNITED STATES.

Harlan...... 8,158	Kimball 959	Nuckolls ...11,417	Saline........20,097	Thomas 517
Hayes........ 3,953	Knox 8,582	Otoe..........25,403	Sarpy........ 6,875	Thurston ... 3,176
Hitchcock.. 5,799	Lancaster...76,395	Pawnee......10,340	Saunders....21,577	Valley........ 7,092
Holt..........13,672	Lincoln......10,441	Perkins....... 4,364	Scott's Bl'ff 1,888	Washingt'n11,869
Hooker...... 426	Logan........ 1,378	Phelps 9,869	Seward......16,140	Wayne........ 6,169
Howard..... 9,430	Loup.......... 1,662	Pierce........ 4,864	Sheridan.... 8,687	Webster....11,210
Jefferson...14,850	McPherson 401	Platte.......15,437	Sherman 6,399	Wheeler 1,683
Johnson....10,333	Madison....13,669	Polk..........10,817	Sioux 2,452	York..........17,279
Kearney..... 9,061	Merrick...... 8,758	Red Will'w 8,837	Stanton...... 4,619	Unorganis'd
Keith......... 2,556	Nance........ 5,773	Richardson17,574	Thayer......12,738	territory.. 695
Keya Paha. 3,920	Nemaha ...12,930	Rock.......... 3,083	Total................1,058,970	

NEVADA.—Area, 112,090 square miles.

Churchill... 703	Esmeralda. 2,148	Lander....... 2,266	Nye.......... 1,290	Storey........ 8,806
Douglas..... 1,551	Eureka...... 3,275	Lincoln...... 2,466	Ormsby..... 4,883	Washoe..... 6,089
Elko......... 4,794	Humboldt. 3,434	Lyon......... 1,987	Roop......... 848	White Pine 1,721
Total...45,761				

NEW HAMPSHIRE.—Area, 9,280 square miles.

Belknap ...20,321	Cheshire...29,579	Grafton....37,217	Merrimack 49,435	Strafford ...38,442
Carroll......18,124	Coos 23,211	Hillsboro...93,247	Rocki'gh'm49,650	Sullivan....17,304
Total..376,530				

NEW JERSEY.—Area, 8,820 square miles.

Atlantic....28,836	Cumberl'd..45,438	Hunterdon 35,355	Morris......54,101	Somerset ...28,311
Bergen.....47,226	Essex......256,098	Mercer......79,978	Ocean......15,974	Sussex......23,259
Burlington.58,528	Gloucester.28,649	Middlesex.61,754	Passaic....105,046	Union.......72,467
Camden....87,687	Hudson...275,126	Monmouth.69,128	Salem......25,151	Warren.....36,553
Cape May..11,268	Total...1,444,933			

NEW YORK.—Area, 47,000 square miles.

Albany....164,555	Dutchess...77,879	Livingston..37,801	Otsego......50,861	Steuben.....81,473
Allegany...43,240	Erie........332,981	Madison....42,892	Putnam 14,849	Suffolk......62,491
Broome.....62,973	Essex........33,052	Monroe....189,586	Queens.....128,059	Sullivan....31,031
Cattaraug's60,866	Franklin ...38,110	Montg'm'y 45,699	Rensselaer124,5 1	Tioga........29,935
Cayuga.....65,302	Fulton........37,650	N. York 1,515,301	Richmond .51,693	Tompkins..32,928
Chautauq'a75,202	Genesee....33,265	Niagara.....62,491	Rockland...35,162	Ulster........87,062
Chemung..48,265	Greene 31,598	Oneida......122,922	S.Lawrence85,048	Warren.....27,866
Chenango..37,776	Hamilton... 4,762	Onondaga.140,247	Saratoga....57,663	Washingt'n45,690
Clinton......46,437	Herkimer..45,608	Ontario.....48,453	Sch'nect'dy29,797	Wayne.......49,729
Columbia..46,172	Jefferson...68.806	Orange......97,859	Schoharie..29,164	W'tch'st'r 146,772
Cortland...25,657	Kings.......838,547	Orleans 30,803	Schuyler ...16,711	Wyoming...31,193
Delaware..45,496	Lewis......29,806	Oswego......71,883	Seneca......28,227	Yates.........21,001
Total..5,997,853				

NORTH CAROLINA.—Area, 50,704 square miles.

Alamance..18,271	Clay.......... 4,197	Guilford....28,052	Montg'm'y.11,239	Rutherford18,770
Alexander. 9,430	Cleveland ..20,394	Halifax....28,908	Moore........20,479	Sampson . .25,096
Alleghany.. 6,523	Columbus..17,856	Harnett 13,700	Nash........20,707	Stanly......12,136
Anson.......20,027	Craven.......20,533	Haywood..13,346	N. Hanover24,026	Stokes.......17,199
Ashe.........15,628	Cumberl'd..27,321	Henderson.12,589	Northamp'n 21,242	Surry........19,281
Beaufort...21,072	Currituck .. 6,747	Hertford. ..13,851	Onslow......10,303	Swain........ 6,577
Bertie.......19,176	Dare........ 3,768	Hyde......... 6,903	Orange.......14,948	Transylv'a. 5,881
Bladen......16,763	Davidson...21,702	Iredell.......25,462	Pamlico..... 7,146	Tyrrell 4,225
Brunswick.10,900	Davie........11,621	Jackson..... 9,512	Pasquot'nk10,748	Union........21,259
Buncombe.35,266	Duplin......18,690	Johnston...27,239	Pender......12,514	Vance.......17,581
Burke.......14,939	Durham ...18,041	Jones........ 7,403	Perquim'ns 9,293	Wake........49,207
Cabarrus...18,142	Edgecombe 24,113	Lenoir......14.879	Person......16,151	Warren......19,360
Caldwell...12,298	Forsyth 28 434	Lincoln......12,586	Pitt...........25,519	Washingt'n10,200
Camden 5,667	Franklin ...21 090	McDowell..10,939	Polk.......... 5,902	Watauga...10,611
Carteret....10,825	Gaston......17,764	Macon 10 102	Randolph...25,195	Wayne......26,100
Caswell....16,028	Gates.........10,252	Madison....17,805	Richmond..23,948	Wilkes.......22,675
Catawba...18,689	Graham..... 3,313	Martin......15,221	Robeson....31,483	Wilson......18,644
Chatham ..25,413	Granville...24,484	Mecklenb'g42,673	Rocki'gh'm25,368	Yadkin......13,790
Cherokee .. 9 976	Greene......10,039	Mitchell....12,807	Rowan......24,123	Yancey...... 9,490
Chowan..... 9,167	Total..1,617,947			

NORTH DAKOTA.—Area, 72,000 square miles.

Alred........ —	Buford....... 808	Dunn......... 159	G'd Forks..18,357	Logan........ 597
Barnes...... 7,045	Burleigh ... 4,252	Eddy......... 1,377	Griggs 2,817	McHenry... 1,558
Benson..... 2,460	Cass.........19,613	Emmons... 1,971	Hettinger... 81	McIntosh.. 8,248
Billings 170	Cavalier.... 6,471	Flannery... 72	Howard..... —	McKenzie.. 3
Bottineau .. 2,893	Church ...... —	Foster....... 1,210	Kidder..... 1,211	McLean..... 860
Bowman.... 6	Dickey...... 5,573	Garfield. ... 33	La Moure .. 8,187	Mercer....... 428

CENSUS OF THE UNITED STATES. 9

Morton...... 4,728	Ramsey..... 4,418	Sheridan.... —	Towner..... 1,450	Ward......... 1,681
Mountraille 122	Ransom..... 5,393	Stark........ 2,304	Traill 10,217	Wells 1,212
Nelson 4,293	Renville.... 99	Steele 3,777	Wallace..... 24	Williams ... 109
Oliver...... 464	Richland...10,751	Stevens...... 16	Wallette ... —	Unorganis'd
Pembina ..14,334	Rolette...... 2,427	Stutsman... 5,266	Walsh 16,587	territory.. 511
Pierce....... 905	Sargent...... 5,076	Total...182,719		

OHIO.—Area, 89,964 square miles.

Adams 26,093	Darke........42,961	Hocking......22,658	Miami 39,754	Sandusky...30,617
Allen.........40,644	Defiance....25,769	Holmes.....21,139	Monroe.....25,175	Scioto........35,377
Ashland ...22,223	Delaware....27,189	Huron......31,949	M'tgome'y100,852	Seneca......40,869
Ashtabula.43,655	Erie...........35,462	Jackson.....28,408	Morgan. ...19,143	Shelby 24,707
Athens.....35,194	Fairfield....33,939	Jefferson. ..39,415	Morrow ...'8,120	Stark........84,170
Auglaize...28,100	Fayette 22,309	Knox.........27,600	Muskingm 51,210	Summit... 54,089
Belmont...57,413	Franklin..124,087	Lake..........18,235	Noble 20,753	Trumbull..42,373
Brown 29,809	Fulton........22,023	Lawrence...39,556	Ottawa 21,974	Tuscarawas46,618
Butler......48,597	Gallia........27,005	Licking 43,279	Paulding ...25,932	Union 22,860
Carroll......17,566	Geauga......13,489	Logan........27,386	Perry 31,151	Van Wert.29,671
Champaign 26,980	Greene.....29,820	Lorain......40,595	Pickaway ..26,959	Vinton 16,045
Clark........52,277	Guernsey...28,645	Lucas 102,296	Pike 17,482	Warren......25,468
Clermont...33,553	Hamilton..374,573	Madison...20,057	Portage.....27,868	Washingt'n42,380
Clinton.......4,240	Hancock ..42,563	Mahoning...55,979	Preble........23,421	Wayne.......39,005
Col'mbiana 59,029	Hardin......28,939	Marion......24,727	Putnam......30,188	Williams ...24,897
Coshocton ..26,703	Harrison ..20,830	Medina....21,742	Richland...38,072	Wood.......44,392
Crawford...31,927	Henry 25,080	Meigs........29,813	Ross 39,454	Wyandot ..21,722
Cuyahoga.309,970	Highland..29,048	Mercer......27,220	Total......................................3,672,316	

OREGON.—Area, 102,606 square miles.

Baker........ 6,764	Curry........ 1,709	Josephine. 4,878	Marion......22,934	Umatilla...13,381
Benton...... 8,650	Douglas....11,864	Klamath... 2,444	Morrow ... 4,205	Union........12,044
Clackamas.15,233	Gilliam..... 3,600	Lake.......... 2,604	Multnomah74,884	Wallowa... 3,661
Clatsop.....10,016	Grant........ 5,080	Lane.........15,198	Polk 7,858	Wasco........ 9,183
Columbia... 5,191	Harney..... 2,559	Linn...........16,265	Sherman ... 1,792	Washingt'n11,972
Coos......... 8,874	Jackson....11,455	Malheur... 2,601	Tillamook...2,932	Yamhill ...10,692
Crook........ 3,244	Total..313,767			

PENNSYLVANIA.—Area, 46,000 square miles.

Adams......33,486	Chester 89,377	Fulton.......10,187	McKean.....46,863	Snyder......17,651
Allegheny551,959	Clarion 36,802	Greene.....28,935	Mercer......55,744	Somerset ...37,317
Armstrong.46,747	Clearfield...69,565	Huntingd'n35,751	Mifflin........19,996	Sullivan....11,620
Beaver.......50,077	Clinton......28,685	Indiana.....42,175	Monroe.....20,111	S'queh'na.40,003
Bedford....38,644	Columbia...36,832	Jefferson...44,005	M'ntg'm'y123,290	Tioga..........52,313
Berks 137,327	Crawford...65,324	Juniata....16,655	Montour...15,645	Union........17,820
Blair 70,866	Cumberl'd..47,271	L'kaw'nna142,088	N'thampt'n84,220	Venango....46,640
Bradford ...59,233	Dauphin...96,977	Lancaster 149,095	Nothumb'd74,698	Warren.....37,585
Bucks.........70,615	Delaware....74,683	Lawrence...37,517	Perry 26,276	Washingt'n71,155
Butler 55,339	Elk 22,239	Lebanon....48,131	Philad'l1,046,964	Wayne.......31,010
Cambria.....66,375	Erie...........86,074	Lehigh......76,631	Pike............ 9,412	Westmo'd.112,819
Cameron ... 7,238	Fayette.....80,006	Luzerne ...201,203	Potter........22,778	Wyoming...15,891
Carbon......38,624	Forest......... 8,482	Lycoming...70,579	Sch'lkill....154,163	York..........99,489
Centre......43,269	Franklin ..51,433	Total...5,258,014		

RHODE ISLAND.—Area 1,806 square miles.

Bristol 11,428	Kent.......26,754	Newport ...28,552	Pr'vid'nce255,123	Washingt'n 23,649
Total...345,506			

SOUTH CAROLINA.—Area, 29,385 square miles.

Abbeville...46,854	Chester 26,660	Florence....25,027	Laurens......31,610	Pickens...16,389
Aiken 31,822	Ch'sterfield18,408	Georget'wn20,857	Lexington ..22,181	Richland...36,821
Anderson ..43,066	Clarendon..23,233	Greenville..44,310	Marion......29,976	Spartanb'g.55,385
Barnwell...44,613	Colleton....40,293	Hampton...20,544	Marlboro...23,500	Sumter.......43,605
Beaufort....34,119	Darlington.20,134	Horry 19,256	Newberry..26,434	Union........25,363
Berkeley ..55,428	Edgefield...49,259	Kershaw.. 22,361	Oconee......18,687	W'msburg.27,777
Charleston.59,903	Fairfield....28,599	Lancaster ..20,761	Orangeb'g..49,393	York..........38,831
Total..1,151,149				

SOUTH DAKOTA.—Area, 78,932 square miles.

Aurora...... 5,045	Buffalo...... 923	Codington. 7,037	Douglas.... 4,600	Hamlin...... 4,625
Beadle 9,586	Butte........ 1,037	Custer....... 4,891	Edmunds.. 4,399	Hand 6,546
BonHomme 9,057	Campbell... 3,510	Davison 5,449	Ewing 16	Hanson 4,267
Boreman... —	CharlesMix 4,178	Day........... 9,168	Fall River. 4,478	Harding.... 167
Brookings.10,132	Choteau 8	Delano 40	Faulk 4,062	Hughes...... 5,044
Brown......16,855	Clark........ 6,728	Deuel......... 4,574	Garfield..... 6,814	Hutchins'n10,469
Brule........ 6,737	Clay......... 7,509	Dewey....... —	Gregory... 295	Hyde......... 1,360

10   CENSUS OF THE UNITED STATES.

Jackson.... 30	Marshall... 4,544	Pratt......... 23	Spink......10,581	Walworth... 2,153
Jerauld.... 3,605	Martin....... 7	Presho..... 181	Stanley.... 1,028	Washaba'h —
Kingsbury. 8,562	Meade...... 4,640	Pyatt....... 34	Sterling.... 96	Washingt'n 40
Lake......... 7,508	Meyer........ —	Rinehart... —	Sully........ 2,412	Yankton....10,444
Lawrence...11,673	Miner........ 5,165	Roberts..... 1,997	Todd......... 188	Ziebach..... 510
Lincoln...... 9,143	Minnehaha 21,879	Rusk.......... —	Tripp......... —	Sisseton &
Lugenbeel.. —	Moody....... 5,941	Sanborn..... 4,610	Turner......10,256	Wahpeton
Lyman...... 233	Nowlin...... 149	Schnasse... —	Union....... 9,130	Indian Re-
McCook.... 6,448	Pennington 6,540	Scobey...... 32	Wagner..... —	servation. —
McPherson 5,940	Potter........ 2,910	Shannon... —	Total.............	328,808

TENNESSEE.—Area, 45,600 square miles.

Anderson...15,125	DeKalb...15,650	Henry.....21,070	Marion......15,411	Sequatchie. 3,027
Bedford....24,739	Dickson....13,645	Hickman...14,499	Marshall...18,906	Sevier.......18,761
Benton....11,230	Dyer........19,878	Houston.... 5,390	Maury......38,112	Shelby....112,740
Bledsoe...... 6,134	Fayette....28,878	Humphr'ys11,720	Meigs........ 6,930	Smith........18,404
Blount......17,589	Fentress..... 5,226	Jackson....13,325	Monroe.....15,329	Stewart......12,193
Bradley....13,607	Franklin...18,929	James........ 4,903	Montg'm'y.29,697	Sullivan....20,879
Campbell...13,486	Gibson.....35,859	Jefferson...16,478	Moore........ 5,975	Sumner....23,668
Cannon....12,197	Giles........34,957	Johnson.... 8,858	Morgan....: 7,639	Tipton......24,271
Carroll.....23,630	Grainger...13,196	Knox........59,557	Obion.......27,273	Trousdale... 5,850
Carter......13,389	Greene.....36,614	Lake.......... 5,304	Overton....12,039	Unicoi....... 4,619
Cheatham.. 8,845	Grundy..... 6,345	Lauderdale18,756	Perry........ 7,785	Union.......11,459
Chester..... 9,069	Hamblen.. 11,418	Lawrence...12,286	Pickett...... 4,736	Van Buren. 2,863
Claiborne..15,103	Hamilton ..53,482	Lewis........ 2,555	Polk......... 8,361	Warren......14,413
Clay......... 7,260	Hancock...10,342	Lincoln.....27,382	Putnam.....13,683	Washingt'n20,354
Cocke.......16,523	Hardeman 21,029	Loudon..... 9,273	Rhea.........12,647	Wayne......11,471
Coffee.......13,827	Hardin......17,698	McMinn....17,890	Roane.......17,418	Weakley....28,955
Crockett ...15,146	Hawkins...22,246	McNairy ...15,510	Robertson..20,078	White........12,348
Cumberl'd.. 5,876	Haywood...23,558	Macon......10,878	Rutherford35,097	Williamson 26,321
Davidson..108,174	Henderson.16,336	Madison....30,497	Scott......... 9,794	Wilson......27,148
Decatur.... 8,995	Total.............			1,767,518

TEXAS.—Area, 287,504 square miles.

Anderson..20,923	Coleman.... 6,088	Gaines........ 68	Jones........ 3,797	Montgom'y11,765
Andrews... 24	Collin......36,736	Galveston ..31,476	Karnes...... 3,637	Moore........ 15
Angelina... 6,306	Collingsw'h 357	Garza........ 14	Kaufman...21,598	Morris....... 6,580
Aransas..... 1,824	Colorado...19,512	Gillespie.... 7,028	Kendall..... 3,809	Motley...... 139
Archer...... 2,101	Comal....... 6,398	Glasscock... 208	Kent:........ 324	Nacogdoc's.15,984
Armstrong. 944	Comanche.16,393	Goliad....... 5,910	Kerr......... 4,445	Navarro ...26,373
Atascosa.... 6,459	Concho..... 1,059	Gonzales...18,016	Kimble..... 2,234	Newton..... 4,650
Austin......17,859	Cooke......24,696	Gray.......... 203	King......... 173	Nolan........ 1,573
Bailey.......—	Coryell.....16,816	Grayson....53,211	Kinney..... 3,781	Nueces...... 8,093
Bandera.... 3,782	Cottle........ 240	Gregg........ 9,402	Knox........ 1,134	Ochiltree... 198
Bastrop....20,736	Crane........ 16	Grimes.....21,312	Lamar......37,302	Oldham.... 270
Baylor...... 2,595	Crockett.... 194	Guadalupe.15,217	Lamb........ 4	Orange..... 4,770
Bee........... 3,720	Crosby...... 346	Hale........ 721	Lampasas.. 7,565	Palo Pinto. 8,320
Bell..........33,297	Dallam...... 75	Hall......... 703	La Salle.... 2,139	Panola......14,328
Bexar.......49,266	Dallas......67,042	Hamilton.. 9,279	Lavaca.....21,887	Parker......21,682
Blanco...... 4,635	Dawson..... 29	Hansford... 133	Lee..........11,952	Parmer...... 7
Borden...... 222	Deaf Smith 179	Hardeman. 3,904	Leon.........13,841	Pecos........ 1,326
Bosque.....14,157	Delta........ 9,117	Hardin...... 3,956	Liberty..... 4,230	Polk........10,332
Bowie......20,267	Denton.....21,289	Harris......37,249	Limestone.21,676	Potter....... 849
Brazoria...11,506	De Witt....14,307	Harrison...26,721	Lipscomb'.. 632	Presidio.... 1,696
Brazos......16,650	Dickens..... 295	Hartley.... 252	Live Oak... 2,055	Rains....... 3,909
Brewster... 710	Dimmit..... 1,049	Haskell..... 1,665	Llano....... 6,759	Randall.... 147
Briscoe...... —	Donley...... 1,056	Hays........11,352	Loving...... 3	Red River..21,452
Brown......11,359	Duval....... 7,598	Hemphill.. 519	Lubbock.... 33	Reeves...... 1,247
Buchel..... 307	Eastland...10,343	Henderson.12,285	Lynn........ 24	Refugio.... 1,239
Burleson...18 001	Ector........ 224	Hidalgo..... 6,534	McCulloch..3,205	Roberts.... 32
Burnet.....10,721	Edwards.... 1,962	Hill.........27,583	McLennan.39,204	Robertson..20,506
Caldwell...15,769	Ellis........31,774	Hockley.... —	McMullen. 1,038	Rockwall... 5,972
Calhoun...* 815	El Paso....15,678	Hood........ 7,581	Madison... 8,512	Runnels.... 3,182
Callahan... 5,434	Encinal..... 1,022	Hopkins...20,572	Marion.....10,862	Rusk........18,559
Cameron...14,424	Erath......21,515	Houston ...19,360	Martin...... 264	Sabine..... 4,909
Camp....... 6,624	Falls........20,706	Howard..... 1,210	Mason...... 5,168	S.August'e. 6,683
Carson...... 356	Fannin....38,709	Hunt.......31,885	Matagorda. 3,985	S. Jacinto.. 7,360
Cass........22,554	Fayette....31,481	Hutchinson 58	Maverick... 3,698	S. Patricio. 1,812
Castro...... 9	Fisher....... 2,996	Irion........ 870	Medina..... 5,730	San Saba... 5,621
Chambers. 2,241	Floyd....... 529	Jack......... 9,740	Menard.... 1,207	Schleicher.. 155
Cherokee..22,975	Foley....... 16	Jackson.... 3,281	Midland... 1,033	Scurry...... 1,415
Childress.. 1,175	Fort Bend.10,586	Jasper...... 5,592	Milam......24,773	Shackelf'rd 2,012
Clay........ 7,503	Franklin... 6,481	Jeff Davis. 1,394	Mills........ 6,430	Shelby 14,365
Cochran.... —	Freestone .15,987	Jefferson... 5,857	Mitchell.... 2,059	Sherman... 71
Coke........ 2,069	Frio.......... 3,112	Johnson...22,313	Montague..18,863	Smith......28,324

CENSUS OF THE UNITED STATES. 11

Somervell. 3,411	Terry 21	Upton 52	Washingt'n29.161	Winkler..... 18
Starr........10,052	Throckm'n 902	Uvalde 3,804	Webb........16 564	Wise 24,134
Stephens ... 4,926	Titus........ 8,190	Val Verde. 2.874	Wharton ... 7,584	Wood.......13,932
Stonewall... 1,024	Tom Green 5,152	Van Zandt.16,225	Wheeler..... 778	Yoakum..... 4
Sutton...... 658	Travis......37,019	Victoria..... 8,737	Wichita.... 4 831	Young 5,049
Swisher..... 100	Trinity...... 7,648	Walker.....12,874	Wilbarger.. 7,192	Zapata 3,562
Tarrant 41,142	Tyler 10,877	Waller 10,888	Williamson25,878	Zavalla...... 1,097
Taylor...... 6,946	Upshur 12,695	Ward....... 77	Wilson 10,655	Total...2,235,523

VERMONT—Area, 10,212 square miles.

Addison...22,277	Crittenden.35,389	Grand Isle. 3,848	Orleans......22,101	Windham..26,547
Bennington20,448	Essex....... 9,511	Lamoille....12,831	Rutland....45,397	Windsor...31,706
Caledonia..23,436	Franklin ...29,755	Orange 19,575	Washingt'n29,606	Total.....332,422

VIRGINIA.—Area, 38,352 square miles.

Accomac ...27,277	Chesterfield26,211	Greensville 8,230	Montgom'y 17,742	Roanoke ...30,101
Albemarle..32,379	Clarke...... 8,071	Halifax......34,424	Nansemond19,692	Rockbridge23,062
Alexandria18,597	Craig........ 3,835	Hanover....17,402	Nelson......15,336	Rocking'm.31,299
Alleghany.. 9,283	Culpeper...13 233	Henrico ...103,394	New Kent.. 5,511	Russell......16,126
Amelia...... 9,068	Cumberl'd 9,482	Henry......18,208	Norfolk.....77,038	Scott........21,694
Amherst....17,551	Dickenson. 5,077	Highland... 5,352	Northam'n.10,313	Shenando'h19,671
Appomatt'x 9,589	Dinwiddie..36,195	Isle of Wi't 11,3 3	Northum'd 7,885	Smyth.......13,360
Augusta....37,005	Elizab'h C'y16,168	James City 5,643	Nottoway...11,582	Southam'n.20,078
Bath........ 4,587	Essex......10 047	Ki'g & Qu'n 9,669	Orange 12,814	Spottsylv'a14,233
Bedford....31,213	Fairfax.....16 655	Ki'g George 6,641	Page 13,092	Stafford ... 7,362
Bland....... 5,129	Fauquier...22 900	K'g Willi'm 9,805	Patrick.....14,147	Surry....... 8,256
Botetourt...14,854	Floyd......14,405	Lancaster... 7,191	Pittsylva'ia59,941	Sussex......11,100
Brunswick.17,245	Fluvanna... 9,508	Lee..........18,216	Powhatan.. 6,701	Tazewell...19,899
Buchanan.. 5,807	Franklin ...24,985	Loudoun...23,274	Pr. Edward14,694	Warren 8,280
B'ckingh'm14,383	Frederick...17,830	Louisa 16,997	Pr. George. 7,872	Warwick ... 6,650
Campbell...41,087	Giles...... 9,090	Lunenburg 11,372	Pr. Anne... 9,510	Washington29,020
Caroline....16,681	Gloucester.11 653	Madison ...10,225	Pr. Willi'm 9,805	Westmor'd. 8,399
Carroll......15,497	Goochland.. 9,958	Mathews... 7,584	Pulaski.....12,790	Wise........ 9,345
Chas. City. 5 066	Grayson ...14,394	Mecklenb'g25,359	Rappahan'k 8,678	Wythe......18,019
Charlotte..15,077	Greene 5,052	Middlesex.. 7,458	Richmond.. 7,146	York........ 7,596
Total..1,655,980				

WASHINGTON.—Area, 69,994 square miles.

Adams...... 2,098	Douglas.... 3,161	Kittitas..... 8,777	Pierce......50,940	Thurston... 9,675
Asotin...... 1,580	Franklin ... 696	Klickitat... 5,167	San Juan... 2,072	Wahkiak'm 2,526
Chehalis... 9,249	Garfield.... 3,897	Lewis......11,499	Skagit...... 8,747	Walla W'la12,224
Clallam.... 2,771	Island 1,787	Lincoln.... 9,312	Skamania.. 774	Whatcom..18,591
Clarke......11,709	Jefferson... 8,368	Mason 2,826	Snohomish. 8,514	Whitman..19,109
Columbia.. 6,709	King.......63,989	Okanogan... 1,467	Spokane.'..37,487	Yakima..... 4,429
Cowlitz..... 5,917	Kitsap...... 4,624	Pacific..... 4,358	Stevens..... 4,341	Total....349,390

WEST VIRGINIA.—Area, 23,000 square miles.

Barbour...12,702	Grant....... 6,802	Logan......11,101	Ohio.......41,557	Taylor......12,147
Berkeley ..18,702	Greenbrier.18 034	McDowell.. 7,300	Pendleton.. 8,711	Tucker...... 6,459
Boone...... 6,885	Hampshire.11,419	Marion....20,721	Pleasants.. 7,539	Tyler.......11,962
Braxton ...13,928	Hancock... 6,414	Marshall...20.735	Pocahontas 6,814	Upshur 12,714
Brooke..... 6,660	Hardy...... 7,567	Mason 22,863	Preston....20,355	Wayne......18,652
Cabell......23,595	Harrison...21,919	Mercer.....16,002	Putnam....14,342	Webster.... 4,783
Calhoun.... 8,155	Jackson...19,021	Mineral....12,085	Raleigh..... 9,597	Wetzel......16,841
Clay....... 4,659	Jefferson ..15,553	Monongalia 15,705	Randolph..11,633	Wirt........ 9,411
Doddridge..12,183	Kanawha..42,756	Monroe....12,429	Ritchie.....16,621	Wood.......28,612
Fayette....20,542	Lewis......15,895	Morgan..... 6,744	Roane.......15,303	Wyoming... 6,247
Gilmer..... 9,746	Lincoln 11,246	Nicholas... 9,309	Summers..13,117	Total......762,794

WISCONSIN.—Area, 58,924 square miles.

Adams...... 6,889	Door........16,682	Kenosha...15,581	Outagamie.38,690	Shawano ...19,236
Ashland...20,063	Douglas....13,468	Kewaunee .16,153	Ozaukee....14,943	Sheboygan.42,489
Barron.....15,416	Dunn.......22,664	La Crosse..38,801	Pepin....... 6,932	Taylor...... 6,731
Bayfield.... 7,390	Eau Claire.30 673	Lafayette...20,265	Pierce......20,385	Trempeal'u18 920
Brown......39,164	Florence... 2,604	Langlade... 9,465	Polk.......12,968	Vernon.....25,111
Buffalo.....15,997	Fond d' Lac44 038	Lincoln....12,008	Portage....24,798	Walworth ..27,860
Burnett.... 4,393	Forest...... 1,012	Manitowoc.37,831	Price....... 5,258	Washburn. 2,926
Calumet....16,639	Grant......36,651	Marathon..30,369	Racine......36,268	Washingt'n22 751
Chippewa..25,143	Green......22,732	Marinette..20,304	Richland ..19,121	Waukesha.33,270
Clark.......17,708	Green Lake15 163	Marquette.. 9,676	Rock........43,220	Waupaca...26,794
Columbia..28,350	Iowa.......22,117	Milwa'kee26,350	St. Croix...30,575	Waushara..18,507
Crawford..15,987	Jackson....15,797	Monroe....23,211	Sauk.......30,575	Winnebago50,097
Dane......59,578	Jefferson ..33,530	Oconto.....15,009	Sawyer..... 1,977	Wood.......18,127
Dodge 44,984	Juneau....17,121	Oneida..... 5,010	Total...1,686,880	

WYOMING.—Area, 97,888 square miles.

Albany	8,865	Converse	2,738	Johnson	2,357	Sheridan	1.972	Uintah	7,881
Big Horn	—	Crook	2,338	Laramie	16,777	Sweetwater	4,941	Weston	2,422
Carbon	6,857	Fremont	2,463	Natrona	1,094	Total			60,705

DISTRICT OF COLUMBIA.—Area, 60 square miles.

The District...230,392

TERRITORIES.

ARIZONA.—Area, 118,916 square miles.

Apache	4,281	Gila	2,021	Maricopa	10,986	Pima	12,673	Yavapai	8,685
Cochise	6,938	Graham	5,670	Mohave	1,444	Pinal	4,251	Yuma	2,671
Total									59,620

NEW MEXICO.—Area, 121,201 square miles.

Bernalillo	20,913	Eddy	—	Mora	10,618	San Miguel	24.204	Socorro	9,595
Chaves	—	Grant	9,657	Rio Arriba	11.534	Santa Fe	13.562	Taos	9,868
Colfax	7,974	Lincoln	7,081	San Juan	1,890	Sierra	3,630	Valencia	13.876
Dona Ana	9,191	Total							153,593

OKLAHOMA.—Area, 2,950 square miles.

Beaver	2,674	Cleveland	6,605	Logan	12,770	Payne	7,215	Greer	5,338
Canadian	7,158	Kingfisher	8,332	Oklahoma	11,742	Total			61,834

UTAH.—Area, 84,476 square miles.

Utah	3,340	Garfield	2,457	Millard	4,033	San Juan	365	Uinta	2,292
Box Elder	7,642	Grand	541	Morgan	1,780	Sanpete	13,146	Utah	23,416
Cache	15,509	Iron	2,683	Pi Ute	2,842	Sevier	6,199	Wasatch	4,627
Davis	6,469	Juab	5,582	Rich	1,527	Summit	7,733	Washingt'n	4,009
Emery	4,806	Kane	1,685	Salt Lake	58,457	Tooele	3,700	Weber	23,065
Total									207,905

CENSUS OF 1890.—Cities and Towns Having a Population of 8,000 and over.

Adams, Mass	9,213	Beatrice, Neb	13,836	Chester, Pa	20,226	E. Portl'd, Ore	10,532
Adrian, Mich	8,756	Beaver Falls, Pa	9,735	Cheyenne, Wyo	11,690	E. Provid'e, R.I.	8,422
Akron, Ohio	27,601	Bellaire, O	9,934	Cincinnati, O	296,908	E. St. Louis, Ill.	15,169
Alameda, Cal	11,165	Belleville, Ill	15,361	Cleveland, O	261,353	Eau Claire, Wis	17,415
Albany, N. Y	94,923	Beverly, Mass	10,821	Clinton, Ia	13,619	Elgin, Ill	17,823
Alexandria, Va	14,339	Biddeford, Me	14,443	Clinton, Mass	10,424	Elizab'hCy, N.J.	37,764
Allegheny, Pa	105,287	Binghamt'n, N.Y.	35,005	Cohoes, N. Y	22,509	Elkhart, Ind	11,360
Allentown, Pa	25,228	Birmingh'm, Ala	26,178	Colo. Sp'gs, Colo	11,140	Elmira, N. Y	29,708
Alpena, Mich	11,283	Bloomington, Ill	20,048	Columbia, Pa	10,599	El Paso, Tex	10,338
Alton, Ill	10,294	Boston, Mass	448,477	Columbia, S. C	15,353	Erie, Pa	40,634
Altoona, Pa	30,337	Braddock, Pa	8,561	Columbus, Ga	17,303	Evansville, Ind	50,756
Amesbury, Mass	9,798	Bradford, Pa	10,514	Columbus, O	88,150	Everett, Mass	11,068
Amsterdam, N.Y.	17,336	Bridgeport, Ct	48,866	Concord, N. H	17,004	Fall River, Mass	74,398
Anderson, Ind	10,741	Bridgeton, N. J	11,424	Corning, N. Y	8,550	Findlay, O	18,553
Ann Arbor, Mich	9,431	Brockton, Mass	27,294	Council B'fs, Ia.	21.474	Fitchburg, Mass	22,037
Anniston, Ala	9,876	Brookline, Mass	12,103	Covington Ky	37,371	Flint Mich	9,803
Appleton, Wis	11,869	Brooklyn, N.Y.	806,343	Cranston, R. I	8,099	Flushing, N. Y	10,868
A'k'ns'sCy, Kan	8,347	Brunswick, Ga	8,459	Cumberl'd, Md	12,729	Fond du Lac, Wis	12,024
Asheville, N. C	10,235	Buffalo, N. Y	255,664	Cumberl'd, R.L.	8,090	Ft. Scott Kans	11,946
Ashland, Wis	9,956	Burlington, Ia	22,565	Dallas, Tex	38,067	Ft. Smith Ark	11,311
Ashtabula, O	8,338	Burlington, N.J.	8,322	Danbury, Ct	16,552	Ft. Wayne, Ind	35,393
Atchison, Kans	13,963	Burlington, Vt	14,590	Danville, Ill	11,491	Ft. Worth, Tex	23,076
Athens, Ga	8,639	Butler, Pa	8,734	Danville, Va	10,305	F'm'gh'm, Mass	9,239
Atlanta, Ga	65,533	Butte, Mont	10,723	Davenport, Ia	26,872	Frederick, Md	8,193
Atlantic Cy, N.J.	13,055	Cairo, Ill	10,324	Dayton, O	61,220	Freeport, Ill	10,189
Auburn, Me	11,250	Cambridge, Mass	70,028	Decatur, Ill	16,841	Fresno, Cal	10,818
Auburn, N. Y	25,858	Camden, N. J	58,313	Delaware, O	8,224	Galesburg, Ill	16,264
Augusta, Ga	33,300	Canton, O	26,189	Denison, Tex	10,958	Galveston Tex	29,084
Augusta, Me	10,527	Carbondale, Pa	10,833	Denver, Colo	106,713	Gardner, Mass	8,424
Aurora, Ill	19,688	Cedar Rapids, Ia	18,020	Des Moines, Ia	50,093	Gloucester, Mass	24,651
Austin, Tex	14,476	Charleston, S.C.	54,955	Detroit Mich	205,876	Glovers'v'le N.Y.	15,864
Baltimore, Md	434,439	Charlotte, N. C	11,557	Dover, N. H	12,790	G'd Rapids, Mich	60,278
Bangor, Me	19,103	Chattan'a, Tenn	29,100	Dubuque Ia	30,311	Green Bay, Wis	9,069
Bath, Me	8,723	Chicago, Ill	1,099,850	Duluth, Minn	33,115	Greenville, S. C.	8,607
Baton Rouge, La	10,478	Chicopee, Mass	14,050	Dunkirk, N. Y	9,416	Greenwich, Ct	10,131
Battle C'k, Mich	13,197	Chillicothe, O	11,288	Dunmore, Pa	8,315	Hagerstown, Md	10,118
Bay City, Mich	27,839	Chip'a F'lls, Wis	8,670	E. Liverpool, O	10,956	Hamilton O	17,565
Bayonne, N. J	19,033	Chelsea, Mass	27,909	Easton, Pa	14,481	Hannibal, Mo	12,857

CENSUS OF THE UNITED STATES. 13

Harrisburg, Pa..39,385
Harrison, N. J... 8,388
Hartford, Ct.....53,230
Hastings, Neb...13.584
Haverhill, Mass27,412
Hazelton, Pa....11,872
Helena, Mont...13,834
Henderson, Ky.. 8,885
Hoboken, N. J..43,648
Holyoke, Mass..35,637
Hornellsv'e,N.Y10,996
Hot Spr'gs Ark. 8 086
Houston, Tex....27,557
Hudson, N. Y... 9,970
H'nt'gton,W.Va 10,108
Hutchinson Kan 6,682
Hyde Park.Mass10,193
Indianap's,Ind 105,436
Ironton, O........10,939
Iron M'nt'n,Mich 8,599
Ishpeming Mich11,197
Ithaca, N. Y....11,079
Jackson, Mich...20,798
Jackson, Tenn...10,039
Jacksonv'e, Fla.17,201
Jacksonv'e. Ill...10,740
Jamestown N.Y.16,038
Janesville. Wis.10,836
Jeffersonv'e Ind.10,666
Jersey Cy, N.J.163,003
Johnston. R. I... 9,778
Johnstown, Pa...21 805
Joliet, Ill.........23 264
Joplin Mo....... 9,943
Kalamazoo,Mich17 853
Kankakee, Ill... 9,025
Kansas Cy, Kan.38,316
Kansas Cy Mo.132,716
Kearney. Neb... 8,074
Keokuk. Ia......14,101
Key West Fla...18.080
Kingston, N.Y.21,261
Knoxville Tenn.22,535
Kokomo, Ind......8,261
La Crosse Wis..25,090
Lafayette. Ind...16,243
Lancaster. Pa...32,011
Lansing, Mich...13,10?
Lansingb'g N.Y.13,550
Laredo, Tex.....11 319
La Salle Ill..... 9,855
Lawrence. Kans. 9,907
Lawrence, Mass.44,654
Leadville. Colo..11,212
Leavenw'h. Kan.19,768
Lebanon. Pa.....14.664
Lewiston. Me...21,701
Lexington. Ky...21,567
Lima, O 15,987
Lincoln, Neb....55,154
Lincoln. R. I....20,355
Little Falls,N.Y. 8.783
Little Rock. Ark25,874
Lockport, N. Y.16,038
Logansport, Ind13,328
L'gIs'd Cy, N.Y.30,500
Los Angeles Cal.50,395
Louisville. Ky..161,129
Lowell, Mass....77,696
Lynchburg, Va..19,709
Lynn, Mass.....55,727
McKeesport, Pa.20,741
Macon, Ga.......22,746

Madison, Ind..... 8,937
Madison, Wis....13,426
Mahanoy, Pa....11,286
Malden Mass....23 031
Manchester. Ct.. 8.222
Manchester,N.H44,126
Manchester, Va. 9,246
Manistee, Mich..12,812
Mankato, Minn.. 8.838
Mansfield, O.....13,473
Marbleh'd. Mass 8,202
Marietta. O....... 8,273
Marinette, Wis..11,523
Marion, Ind...... 8,769
Marion, O........ 8,327
Marlboro. Mass.13,805
Marquette, Mich 9 093
Marshalltown Ia 8,9 4
Massillon, O.....10,092
Meadville, Pa... 9,020
Medford, Mass..11,079
Melrose, Mass... 8,519
Memphis,.Tenn.64,495
Men'min'e, Mich10,630
Meriden, Ct.....21,652
Meridian, Miss..10,624
Michi'n Cy, Ind.10,776
Middletown, Ct.. 9 013
Middlet'n, N. Y. 11,977
Milford, Mass... 8,780
Millville, N. J...10,002
Milwa'kee. Wis204,468
Minnea's, Minn164,738
Moberly, Mo.... 8,215
Mobile, Ala......31,076
Moline, Ill......12,000
Montg'm'ry. Ala21,883
Mt. Carmel, Pa.. 8,254
Mt. Vernon,N.Y10,677
Muncie. Ind.....11,345
Muscatine. Ia...11,454
Muskegon,Mich.22,702
Nanticoke, Pa...10,044
Nashua. N. H... 19,311
Nashville Tenn.76,168
Natchez, Miss...10,101
Natick, Mass.... 9,118
Nebraska C'y Neb11,494
N. Albany. Ind..21,059
Newark,N. J...181,830
Newark, O......14,270
N. Bedf'd. Mass 40,733
N. Bright'n.N.Y16,423
N. Britain, Ct...19,007
N. Br'nsw'k,N.J18,603
Newburg. N. Y..23,087
N'wb'ryp't.Mass13,947
Newcastle. Pa...11,600
New Haven, Ct..81.298
New London. Ct13,757
N'w Orleans La242,039
Newport, Ky....20,918
Newport. R. I...19,457
N. Rochelle,N.Y 8,818
Newton. Mass...24,379
N. York, N.Y.1,515.301
Norfolk. Va.....34,871
Norristown. Pa..19.791
N'h Adams,Mass16.074
Northa'n, Mass.14 990
Norwalk, Ct.....17,747
Norwich. Ct....16.156
Oakland, Cal...48,082

Ogden, Utah......14.889
Ogdensb'g, N.Y.11,662
Oil City, Pa....10,932
Omaha, Neb....140,452
Orange. N. J....18,844
Oshkosh, Wis ..22,836
Oswego, N. Y....21,842
Ottawa City, Ill.. 9,985
Ottumwa Ia.....14,001
Owensboro. Ky.. 9,837
Paducah. Ky....13,076
Paris, Tex....... 8,254
Parkers'g,W.Va 8,408
Passaic, N. J....13,028
Paterson. N. J..78.347
Pawtucket. R. I.27,633
Peabody, Mass..10,158
Peekskill, N. Y. 9,676
Pensacola, Fla..11,750
Peoria Ill........41,024
P'th Amboy,N J 9,512
Petersburg. Va..22,680
Philad'a, Pa..1,046,964
Phillipsb'g, N.J. 8,644
Phœnixvi'e, Pa.. 8.514
Pine Bluff, Ark.. 9,952
Piqua, O......... 9,090
Pittsburg Pa... 38,617
Pittsfield. Mass.17.281
Pittston, Pa....10.302
Plainfield, N. J.11,267
Plattsm'th. Neb. 8,392
Plymouth, Pa... 9 344
P't Huron, Mich13 543
P't Jervis N. Y. 9 327
Portland, Me....36 425
Portland Ore ..46,385
Portsm'th N. H. 9,827
Portsm'th, O....12,394
Portsm'th. Va...11,417
Pottstown, Pa ..13,285
Pottsville. Pa...14,117
Po'keepsie, N.Y.22.206
Providc'e. R. I.132,146
Pueblo, Colo....24,558
Quincy, Ill......31.494
Quincy, Mass...16,7-3
Racine, Wis.....21,014
Raleigh, N. C...12,678
Reading, Pa.....58,661
Richmond, Ind.16,608
Richmond, Va..81,388
Roanoke, N. C..16,159
Rochester,N.Y.133,896
Rockford, Ill....23,584
Rock Island. Ill.13,634
Rockland, Me... 8,174
Rome, N. Y.....14,991
Rutland, Vt.....11,760
Sacramento Cal.:6,386
Saginaw, Mich..46,322
St. Joseph, Mo.52,324
St. Louis, Mo...451,770
St. Paul, Minn.133,156
Salem, Mass.....30,801
S't L'e Cy, Utah.44,843
S. Antonio. Tex.37,673
San Diego, Cal..16,159
Sandusky. O....18,471
S. Franc'o, Cal.298,997
San Jose, Cal...18,060
Sar'a Sp's, N.Y.11,975
Savannah, Ga..43,189

S henec'y. N.Y..19.902
Scranton. Pa.....75.215
Seattle, Wash ..42,837
Sedalia, Mo 14,068
Shamokin, Pa...14,403
Sheboygan,Wis.16,359
Shenandoah,Pa.15,944
Shreveport, La..11,979
Sing Sing, N. Y. 9,352
Sioux City, Ia..37,806
Sioux F's. S.Dak10,177
Somerville,Mass40,152
South Bend,Ind 21,819
Stillw'r, Minn..11,260
S. Bethl'm, Pa..10,302
S. Omaha. Neb.. 8,062
Spencer, Mass... 8,747
Spok's F's,Wash19,922
Springfield, Ill..24,963
Springf'd ,Mass 44,179
Springfield, Mo.21,850
Springfield, O...31,895
Stamford, Ct....15,700
Steelton, Pa..... 9,250
Steubenville, O.13,394
Stockton, Cal...14,424
Streator, Ill.....11,414
Superior, Wis...11,983
Syracuse, N. Y..88,143
Tacoma, Wash..36,006
Taunton, Mass..25,448
Terre Haute Ind30,217
Tiffin, O.........10,801
Titusville, Pa.... 8,073
Toledo, O.........81.434
Topeka. Kans...31,007
Trenton, N. J...57,468
Troy, N. Y......60,956
Union, N. J.....10,643
Utica, N. Y.....44,007
Vernon, Ct...... 8,808
Vicksburg, Miss 13,373
Vincennes, Ind. 8,853
Waco, Tex......14,445
Waltham, Mass.18,707
Warwick R. I...17,761
Washing'n,D,C 230.392
Waterbury, Ct..28,646
Watert'wn.N.Y.14,725
Watert'wn. Wis. 8,755
Wausau, Wis.... 9,253
W.B'y Cy, Mich.12,981
W. Chester, Pa.. 8,028
Westfield. Mass. 9,805
W. Troy, N. Y..12,967
Weymo'h, Mass.10,866
Wheel'g, W.Va.35,013
Wichita, Kans...23,853
Wilkesbarre.Pa.37,718
W'msport, Pa...27,132
Williamnti. Ct. 8,648
Wilmington,Del 61,431
Wilming'n. N.C.20,056
Winona, Minn..18,208
Winston, N. C. 8,018
Woburn, Mass..13,499
Woonsoc't. R. I..20,830
Worces'r, Mass..84,655
Yonkers. N. Y...2,088
York, Pa........20,793
Youngstown, O..33,220
Zanesville, O...21,009

THE CAPITOL OF THE UNITED STATES.

THE corner-stone of the Capitol was laid by the illustrious Washington, on the 18th day of September, 1793. The building was opened for the meeting of Congress November 17th, 1800. Enlargement and new dome completed in 1867. The edifice fronts the east, is 751 feet long, 348 feet wide, and covers 3½ acres; courtyards, 3½ acres; in all 7 acres. The predominant material of the exterior is white marble. The dome is of cast-iron, 135½ feet in largest diameter, and 287½ feet high, surmounted by a statue of Liberty 19½ feet high. The interior of the dome forms a remarkable circular chamber, or rotunda, 96 feet in diameter, 180 feet high. One thousand gas jets, flashed by electricity, illuminate the interior by night. The walls of the rotunda are adorned with historical paintings by eminent artists. The Senate Chamber, House of Representatives, Supreme Court Rooms, and other apartments are splendidly decorated. The halls are lined with polished marbles from every State in the Union. Frescoes, paintings, and sculptures abound. The front porticoes are supported by one hundred Corinthian columns of white marble. The cost of the Capitol building was thirteen millions of dollars. It may be justly styled the PALACE OF LAWS, for within its precincts the statutes of the nation are enacted and expounded. Here are framed the patent laws and kindred ordinances for the encouragement of authors and inventors.

THE UNITED STATES.

THE greatest length from East to West is 2800 miles; greatest breadth North to South, 1600 miles; average breadth, 1200 miles. Total area, 3,026,494 square miles; area of Alaska additional, 577,390 square miles. The shores of the Atlantic are for the most part low, those of the Pacific rocky. The length of the Atlantic coast line is 2349 miles; Gulf, 1556; Pacific, 1810, indentations not included. Shore lines of the great lakes, 3450 miles. Number of States, 44. Length of the Mississippi River, 2900 miles; Missouri, 3000 miles. Length of steam navigation, Mississippi River and branches, 7100 miles; ditto Missouri River, 3000 miles; ditto Ohio River, 3292 miles; ditto Red River, 3630 miles. Approximate total length of railways in operation in United States, 1891, 175,000 miles.

THE PATENT LAWS.

WITH DIRECTIONS AND COSTS FOR OBTAINING PATENTS, CAVEATS, TRADE-MARKS, COPYRIGHTS, ETC., WITH ABSTRACTS FROM OFFICIAL RULES.

IN the practical application of new and useful improvements, America leads the world; according to an estimate made by the Commissioner of Patents, from six to seven eighths of the entire manufacturing capital of the United States, or upward of six thousand millions of dollars, probably, is based upon patents, either directly or indirectly. A very large proportion of all patents prove remunerative; which is the reason why so many are applied for, and so many millions of capital invested in their working. "But all patents," says an able writer, "are not productive; neither are all farms; all men are not rich; all mines are not bonanzas.

"There is scarcely an article of human convenience or necessity in the market to-day, that has not at some time or other been the subject of a patent, either in whole or in part. The sale of every such article yields the inventor a profit. If we purchase a box of paper collars, a portion of the price goes to the inventor; if we buy a sewing-machine, the chances are that we pay a royalty to as many as a dozen or fifteen inventors at once. Indeed, the field is so vast and the number of profitable patents so great that it would be far preferable to undertake a recapitulation of those patents which are not profitable than those which are."

HOW MUCH IS A PATENT WORTH?

IN an official report, a chief examiner of the Patent Office says: "A patent, if it is worth any thing, when properly managed, is worth and can easily be sold for from ten to fifty thousand dollars. These remarks only apply to patents of ordinary or minor value. They do not include such as the telegraph, the planing machine, and the rubber patents, which are worth millions each. A few cases of the first kind will better illustrate my meaning:

"A man obtained a patent for a slight improvement in straw-cutters, took a model of his invention through the Western States, and after a tour of eight months returned with forty thousand dollars in cash, or its equivalent.

"Another inventor obtained extension of a patent for a machine to thresh and clean grain, and sold it in about fifteen months for sixty thousand dollars. A third obtained a patent for a printing-ink, and refused fifty thousand dollars, and finally sold it for about sixty thousand dollars.

"These are ordinary cases of minor invention, embracing no very considerable inventive powers, and of which hundreds go out from the Patent Office every year. Experience shows that the most profitable patents are those which contain very little real invention, and are to a superficial observer of little value."

THE PATENT OFFICE AT WASHINGTON.

THE engraving on page 24 shows a full exterior view of the Patent Office, which is one of the finest edifices in Washington. It is of the Doric order of architecture, 433 feet long, 331 feet wide, 75 feet high. The collection of models of inventions here gathered is very remarkable, the aggregate number being over two hundred thousand. Nearly forty thousand new applications for patents are sent to the Patent Office each year.

PROCEEDINGS TO OBTAIN A PATENT.

To one who has made an invention or discovery, the first inquiry that suggests itself is, "Can I obtain a Patent?" If so, "How shall I proceed? Whom shall I consult? How much will it cost?" The quickest way to settle these queries without expense is to write to us (Munn & Co.) describing the invention. Send us also a small sketch. Never mind your inexperience. Nicety of writing or drawing is not essential; all we need is to get *your idea*. Do not use pale ink. Be brief. Send stamps for postage. We will immediately answer and inform you whether or not your improvement is probably patentable; and if so, give you the necessary instructions for further procedure. Our long experience enables us to decide quickly. For this advice we make *no charge*. All who desire to consult us in regard to obtaining patents, are cordially invited to do so. We shall be happy to see them in person at our office, or to advise them by letter. In all cases they may expect from us a careful consideration of their plans, an honest opinion, and a prompt reply.

Inquiries about the *patentability of new inventions* we answer, as above stated, *without charge*. But we frequently receive letters containing strings of other questions, without fee, or even postage-stamps. For this class of inquirers the following hints may be useful: The best washing-machines, the best brick machines, *the best of everything in the mechanical line*, is advertised and illustrated in the SCIENTIFIC AMERICAN, and the address of the parties having such things on sale is there given. If not a subscriber to the SCIENTIFIC AMERICAN, you should enroll your name by sending $3.00 for one year, which includes the postage. You will sooner or later find in its pages answers to all your inquiries, together with an immense amount of other useful information.

THE PRELIMINARY EXAMINATION.

THIS consists of a *special search*, made among the records of the patents that have been granted, to ascertain whether any invention is noticed that will probably prevent the grant of a patent to our client. On the completion of this special search, we send a *written report* to the party concerned, with suitable advice. Our charge for this service is five dollars.

In making this search, we do not guarantee that none of the patents will be overlooked, as the number is enormous; and we do not guarantee that a patent will be granted, even if the Preliminary Report is favorable. But in general, if the report is unfavorable, the applicant will be saved all further expense; while, if favorable, he will generally, but not always, be enabled to obtain a patent.

In offering to make this examination for five dollars, our correspondents must bear in mind that we here refer only to the question of the *patentability* of the invention, not to infringements or other questions. Will it pay? Does it infringe? See page 40 for reply.

All that we need for a preliminary examination is a brief description and sketch sufficient to enable us to get an idea of the invention.

The fee paid for preliminary examination does not go toward paying for the patent.

The most prudent way for the applicant is to order a preliminary examination. It adds a little to the expenses, but is, on the whole, generally satisfactory. This examination we can quickly make, as we employ, for the purpose, a corps of experienced examiners.

For Preliminary Examinations, send sketch and brief description of the invention to MUNN & CO., 361 Broadway, New York.

OF THE UNITED STATES.

MAP OF THE UNITED STATES.

What security have I that my communications to Munn & Co. will be faithfully guarded and remain confidential?

Answer.—You have none except our well-known integrity in this respect, based upon a most extensive practice of forty years' standing. Our clients are numbered by hundreds of thousands. They are to be found in every town and city of the Union. Please to make inquiry about us. Such a thing as the betrayal of a client's interests, when committed to our professional care, never has occurred, and is not likely to occur. All business and communications intrusted to us are kept *secret and confidential.*

HOW TO APPLY FOR A PATENT.

AN application for a patent consists of a Petition, Affidavit of Invention, Drawings and Specifications, all of which must be prepared in legal form and in accordance with official rules.

In order to apply for a patent, all that is necessary is to send either a clear sketch or a model of the invention to MUNN & CO., by express or by mail, prepaid, with an explanation of the merits and working of the invention. Be very particular to give your ideas in full about the invention. Describe its intended working, and mention all the advantages you can think of. This description is always of assistance to us in preparing the specification and drawings. Also remit $25 on account, and give the inventor's full name, middle name included. We will then prepare the above official papers, and send them to you for examination and signature, with directions about signing and verifying the same. The next fee (which is $30 if the case is an ordinary one) is then payable.

The case is then filed in the Patent Office, and in due time receives official examination. When the patent is allowed, the applicant pays $20 more, making the total cost of the patent $75, of which the Government fees are $35 and our (MUNN & CO.S) charges $40 for the drawings, specification and attention to the business of the case before the Patent Office. These costs relate to ordinary cases. When the invention is complicated, additional time and labor are required, and the costs are increased; but we always aim to make them quite moderate.

As soon as the case is filed in the Patent Office, the applicant is protected against the grant, without his knowledge, of a patent for the same thing to another person.

When the Patent Office decides to grant a patent, we send notice to the applicant, stating the application has

THE OFFICIAL EXAMINATION.

been *allowed*, and the patent will be printed and issued as soon as the final Government fee of $20 is paid in. The applicant may pay this at once, and have the patent immediately issued, or he may wait six months before making the payment; meantime, the application will be held in the secret archives of the Patent Office, ready for issue on payment of the final fee. The usual way, if it is contemplated to take foreign patents, is to let the allowed case rest in the secret archives until the foreign applications have been prepared and filed.

When the patent issues, we publish a special notice thereof in the SCIENTIFIC AMERICAN, briefly descriptive of the leading features and merits of the invention, with the patentee's name and address. We print and distribute about *forty thousand copies of this notice*, free of charge to our client. This publication is of value in assisting to bring the invention before the public and promoting its introduction. Were the inventor to do this printing himself, say on the backs of postal cards, it would cost him four hundred dollars for the cards and additional expense for presswork and addressing. A postal card or printed circular is rarely seen by more than one person, whereas a notice in the SCIENTIFIC AMERICAN probably comes before half a million readers.

We have a branch house in Washington (see engraving, page 24), employing a corps of skilled assistants, and we make it our special duty to watch over the cases of our clients while they are before the Patent Office.

After the application is filed, it receives, in its due turn, an official examination, when the Patent Office examiner makes such objections and cites such references to other patents as he thinks proper. We then examine the references, and use our best endeavors, by written and oral argument, to remove the objections and procure the allowance of the patent. On the second hearing, new objections and new references are often cited, and further time and labor is then required on our part; and so on perhaps for a third or a fourth hearing. It will thus be seen that the work of attending to the case while before the Patent Office is very laborious and consumes much time; the cost therefor is included in our fees already mentioned; and we make no additional charges, except in cases of peculiar difficulty and on advice with the applicant.

The time required to procure a decision from the Patent Office is from six to eight weeks. But in some classes of invention there is an accumulation of work, and it is several months before the case is reached.

WHO MAY TAKE PATENTS.

American patents are granted for the term of 17 years, and cannot be extended. The patent remains good whether the invention is worked or not; and no additional taxes or payments are required beyond the costs on first taking out the patent.

Citizens, foreigners, women, minors and the administrators of estates of deceased inventors, may obtain patents. There is no distinction as to nativity, person or charges.

Two or more persons may apply jointly for a patent, if they are joint inventors. Where one person is the inventor and the other only a partner, the patent must be applied for in the name of the inventor; but he may secure his partner in advance by executing a deed of conveyance, so drawn that the patent will be issued in both names.

An inventor may grant a license, or sell and assign any portion of his right in an invention, either before or after the patent is granted. The deed of conveyance should be recorded in the Patent Office. Our charge (MUNN & CO.s) to prepare a patent deed and attend to the recording of the same is $5.

New medicines or compounds, and useful mixtures, recipes, etc., may be patented. A minute statement must be given of the exact proportions, method, and ingredients used in making a given quantity of the new article.

If model is not furnished, then send us photographs and sketches of the invention, showing side view, front, top or plan views, and sectional elevations, in order that the exact position and working of all the parts may be clearly understood. The full details of arrangement or construction should be given, also the full name of the inventor.

Models are not now required by the Patent Office, except in special cases; but a small model is always useful to the solicitor in preparing the drawings and specification, and in general we recommend the inventor to furnish a model, which we will return after the drawings are prepared.

If a model is sent, it may be quite small and cheaply made, of any convenient materials; whittled out of wood will often answer. All that is necessary is to illustrate the invention. Do not waste time or money in making a large model. The smaller the size the better.

If sent by mail, do *not* nail or screw the box containing the model, but tie it. Do not put any writing in the box or on the model. Violation of these rules compels the Post Office authorities at New York to collect letter postage, or two cents an ounce, before delivery.

In forwarding the model, never place money in the box therewith, as it is liable to be stolen. Remit the money by postal order, check, or draft, to order of MUNN & Co. Send the model by express, prepaid, addressed MUNN & Co., 361 Broadway, New York.

SCIENTIFIC BOOKS.

MESSRS. MUNN & Co. have an extensive book department for the supply of all kinds of scientific, mechanical and other books. Orders promptly filled.

Carpentry Made Easy, or the Science Art of Framing, with Instructions for Balloon Frames, Barn Frames, Mill Frames, Warehouses, Church Spires, etc.; also Bridge Building, with Estimates and Tables. 44 plates and near 200 figures. By William E. Bell. 1888. Price $5.

Trade Secrets and Recipes.—A Collection of Recipes, Processes and Formulæ that have been offered for sale at prices varying from 25 cents to $500. With Notes and Additions. By John Phin. Cloth. Price 60 cents.

Address MUNN & Co., 361 Broadway, New York. A catalogue of books sent free.

GOING TO WASHINGTON.

SOME inventors suppose, very naturally, that if personally present in Washington, they can get their cases through more expeditiously, or command other important facilities. But this is not so. The journey to Washington is usually a mere waste of time and money; but, notwithstanding, some persons prefer to go. A good agent must be employed after the inventor gets there. No one can possibly have facilities or influence superior to our own; a very large portion of the entire business of the Patent Office passes through our hands; and we have an office in Washington, charged with the especial duty of watching over and pressing forward the interests of our clients.

The Patent Office does not prepare patent papers, or make models. These must be provided by the applicant or his attorney, according to law, otherwise his claim will not be considered.

The law especially requires that all documents deposited in the Patent Office shall be correctly, legibly, and clearly written, and that the drawings shall be of a specified size, and executed in an artistic manner.

Persons who visit Washington in person can have all their patent business promptly attended to, by calling at MUNN & CO.'S BRANCH SCIENTIFIC AMERICAN OFFICE, 622 F. st.,

THE UNITED STATES PATENT OFFICE, AT WASHINGTON.—(See page 16.)

Pacific Building, near Seventh street and the Patent Office. (See engraving opposite page.)

REJECTED OR DEFECTIVE CASES.

WE (Munn & Co.) give prompt attention to the prosecution of rejected or postponed cases, that have been prepared by the applicant or other agent. Terms very moderate.

CAVEATS.

THE filing of a Caveat is sometimes of great importance, as it may be quickly done, and affords *immediate protection* against the issue of a patent, without the knowledge of the Caveator, to any other person for the same invention. The object of a Caveat is to give the inventor time to test and perfect his discovery. Should a competitor apply for a patent for the same invention, the Caveator is officially notified, and called upon to file in his application for a patent. The existence of a Caveat is one of the evidences of priority of invention. A Caveat runs for a year, and can be extended from year to year. Caveats can only be filed by citizens of the United States, and aliens who have resided here one year and have declared their intention to become citizens. All Caveats are secret. No one can see or obtain a copy of a Caveat without the order of the Caveator. A Caveator can use the stamp, "Caveat filed;" and such stamp sometimes assists in selling an article, or securing trade.

But the filing of a Caveat does not secure any *exclusive* right of sale. The Patent secures that right. The filing of a Caveat has nothing to do with the grant of a patent. The Government makes no search as to novelty when a Caveat is filed. No portion of the money paid for a Caveat applies toward the patent.

A Caveat consists of a Petition, Specification, Drawing, and Affidavit of Invention. To be of any value, these papers should be carefully drawn up, and the official rules scrupulously complied with. No model is required. Our facilities enable us to prepare Caveat papers with great dispatch. When specially desired, we can have them ready to send to the applicant, for signature and affidavit, by return mail, or at an hour's notice. The whole expense to file a Caveat is generally $30, of which the official fee is $10, and we generally charge $20 to prepare the papers and attend to the business. On filing the Caveat in Washington the Patent Office issues an Official Certificate thereof, which we forward to the applicant. To enable us to prepare Caveat papers, all that we need is a sketch, drawing, or photograph, and description of the invention, with which remit fees as above. Model not required.

APPEALS.

WHEN the examiner refuses to allow a patent, and finally rejects the case, we report the fact to our client, and inform him as to the probabilities of obtaining a reversal of the examiner's decision by appeal.

Three appeals are allowed, namely: to the Examiners-in-Chief, to the Commissioner of Patents, to the Supreme Court of the District.

First Appeal.—The Government fee payable by the applicant, on making an appeal to the Examiners-in-Chief, is $10. Our charges for preparing and conducting this appeal are very moderate, and in part contingent upon success.

Second Appeal.—From the decision of the Examiners-in-Chief an appeal may be taken to the Commissioner of Patents. Government fee, $20.

Third Appeal.—From the decision of the Commissioner of Patents an appeal may be taken to the Supreme Court of the District of Columbia.

REISSUE OF PATENTS.

WHENEVER any mistake, defect, or insufficiency in the claims or specification of a patent are found to exist, a petition for a reissue may be filed in the Patent Office, together with new drawings and corrected specifications. A new corrected patent will then be issued, and the old patent cancelled. Messrs. Munn & Co. have had forty years' experience in obtaining reissues, and will be happy to give further information upon the subject, by letter, to all who wish to have their patents corrected.

INFRINGEMENTS.

THE general rule of law is, that the first original patentee is entitled to a broad interpretation of his claims. The scope of any patent is therefore governed by the inventions of prior date. To determine whether the use of a patent is an infringement of another generally requires a most careful examination of all analogous prior patents and rejected applications. An opinion based upon such research requires for its preparation much time and labor. The expense of these examinations, with written opinion, varies from $25 to $100 or more, according to the labor involved. Address MUNN & CO., 361 Broadway, New York.

Infringements occur much less frequently than most people suppose; and in general, unless you have special reason to believe that infringement exists, the best way is not to give yourself trouble about it until some one troubles you. Infringement consists in the use, sale, or manufacture of something already patented. It is not an infringement to take out a patent for an invention which is an improvement on a previous patent. It is not an infringement to own, to buy, or to sell any patent. It is not an infringement to sell rights under any patent, whether town, county, or State rights, or licenses.

All good improvements are worth patenting, even if their use should be found to infringe a prior patent. Only a few, comparatively, of the large number of patents issued prove to infringe; and the infringing device is sometimes worth more than the patent with which it conflicts. Patentees of conflicting inventions can usually make satisfactory arrangements with the owners of the prior patents; it is obviously to the interest of prior patentees to have their patents used as extensively as possible. The princely revenue of Howe, the inventor of the sewing-machine, was about $500,000 annually, derived chiefly from two infringing patentees, paying him a small royalty on each machine. The net profits divided among the owners of one of these infringing patents—the celebrated Wheeler & Wilson—are reported to be more than $1,000,000 a year. The profits of the other, the Singer Manufacturing Co., are reported at from $2,000,000 to $3,000,000 a year.

ASSIGNMENTS OF PATENTS.

IF you desire to have an assignment of a patent, or any share thereof, or a license, made out in the proper manner, and placed on record, remit five dollars, give full names of parties, residences, title of invention, date of patent. The above charge includes the recording fee.

Inventions or shares thereof may be assigned either before or after the grant of a patent. All transfers or assignments of any share in a patent should be recorded at Washington.

If you desire to know in whose name the title to a patent is officially recorded at Washington; or if you wish for an abstract of all the deeds of transfer connected with a patent, send us the name of the patentee, or date or number of patent, and remit five dollars.

We (MUNN & Co.) have branch offices in Washington, and have constant access to all the public records. We can therefore make for you *any kind of search*, or look up for you *any sort of information* in regard to patents, or inventions, or applications for patents, either pending or rejected, that you may desire.

PATENTS FOR ORNAMENTAL DESIGNS.

THE laws for the grant of patents for new designs are of the most liberal and comprehensive character, and their benefits may be enjoyed by all persons, without distinction as to nationality.

Foreign designers and manufacturers who send goods to this country may secure patents here upon their new patterns, and thus prevent other makers from selling similar goods in this market.

A patent for a design may be granted to any person, whether citizen or alien, who, by his own industry, genius, efforts, and expense, has invented or produced any new and original design for a manufacture, bust, statue, alto-relievo, or bass-relief; any new and original design for the printing of woollen, silk, cotton, or other fabrics; any new and original impression, ornament, pattern, print, or picture, to be printed, painted, cast, or otherwise placed on or worked into any article of manufacture; or any new, useful, and original shape or configuration of any article of manufacture, the same not having been known or used by others before his invention or production thereof, or patented or described in any printed publication, upon payment of the duty required by law, and other due proceedings had the same as in cases of inventions or discoveries.

Patents for designs are granted for the term of three and one half years, or for the term of seven years, or for the term of fourteen years, as the said applicant may elect in his application. The patent expires at the end of the term for which it is first granted. No extension.

Design patents are granted for any new shape, form, or curve given to a tool, or the frame or special part of a machine, and for any configuration that makes an article look better or more desirable to the eye of the purchaser. Thus the scope of the design patent law is very broad. The patentee of a machine may, in addition to the protection of an ordinary patent, also obtain a design patent upon any new ornaments or ornamental forms used on his device.

The personal presence of the applicant is not necessary in order to obtain a design patent, as the business can be done by correspondence. Those who reside at a distance should send us their names in full, middle name included, together with twelve photographs of the design not mounted. Also remit the fees as above, by draft, check, or postal order. We will then prepare the petition, oath, and specification, and forward the same to applicant for signature. On their return by him the papers are filed at the Patent Office, when an official examination is made, and if no conflicting design is found to exist, a patent is issued. The photographs only need to be large enough to represent clearly all the features of the design.

The petition, oath, specification, assignments, and other proceedings in the case of applications for letters-patent for a design are the same as for other patents.

City residents, by calling at our office, can have all the business promptly attended to.

The expenses for design patents are as follows:
Patent for three and a half years, whole expense, $30.
Patent for seven years, whole expense, $35.
Patent for fourteen years, whole expense, $50.
The above includes Government fees and agents' charges.*
Address MUNN & CO., 361 Broadway, N. Y.

COPIES OF PATENTS.

A PRINTED copy of the full specification, with the drawings, of any patent granted since January 1, 1866, is furnished by MUNN & CO. for 25 cents.

A printed copy of the drawing only of any patent granted prior to January 1, 1866, is furnished by MUNN & CO. for 25 cents. The specifications of patents dated prior to 1866 have not yet been printed; but MUNN & CO. will supply written copies thereof at reasonable cost.

A printed copy of the drawing, with a copy of the claim of any patent of date prior to January 1, 1866, is furnished by MUNN & CO. for one dollar.

In order to save the costs of searching, parties who order copies of patents as above should give the date of the patent and the patentee's name. Send the money with the order, and address MUNN & CO., SCIENTIFIC AMERICAN OFFICE, 361 Broadway, New York.

☞ If the patentee's name and date of the patent are unknown, we will, if desired, carefully search for the patent described in the order. For the time occupied in this search we make a reasonable charge.

* The Government fee is $10 for three and a half years, $15 for seven years, and $30 for fourteen years. Our (MUNN & CO.'s) charges are $20. When it is inconvenient for applicants to furnish their own drawings or photographs, we can supply them at a reasonable cost.

TRADE-MARKS.

THE patent law provides that any person, firm, or corporation may secure an exclusive right to use a trade-mark, by complying with the official regulations of the Patent Office. The whole expense is forty-five dollars.

Trade-marks already in use, no matter for how long a time, may be registered; provided the mark is used in commerce with any foreign country or with an Indian tribe.

A trade-mark consists of a distinctive or special name or title for an article, or a device, design, or stamp, or combination thereof, applied to merchandise or the envelopes or packages thereof. But the mere business name of a person or firm is not registerable as a trade-mark.

The official rules must be carefully observed. A petition is to be signed by the applicant, together with a written description of the trade-mark, statement, and declaration as to use, affidavit thereto; a copy of the trade-mark is to be furnished, drawn or mounted on drawing-paper, with twelve copies not mounted.

Trade-marks remain in force for thirty years, and may be renewed for thirty years more. It is unlawful for any person to use any registered trade-mark, or to make such a resemblance thereof as is calculated to deceive. But a trade-mark registered for use upon one particular class of merchandise—hardware goods, for example—will not prevent registration of a similar mark, by another person, for use upon an entirely different class of merchandise—crockery goods, for example.

Those who desire to secure protection for trade-marks are requested to communicate with MUNN & CO., No. 361 Broadway, New York, who make it a part of their business to prepare the papers and attend to the application before the Patent Office.

All the business is speedily done. Registration is generally granted within ten days after the papers are filed. City residents should call at our office. Those who live at a distance should give us, in a letter, the following information:

1. The names of the parties who own the trade-mark, their residence and place of business.

2. State the class of merchandise and the particular description of goods in connection with which the trade-mark is to be used.

3. Describe the particular mode in which the trade-mark has been and is intended to be applied and used. For example, for a trade-mark for sheetings the statement would be, "The trade-mark is to be printed in blue ink upon the

TRADE-MARKS. 31

outside of each piece of sheeting." Or, "The trade-mark is to be printed in black, or red, white, and blue, upon the exterior of a paper wrapper, which is to cover or extend around each package of the goods." In the use of a trade-mark the owner is not confined to such particular colors or precise method of use; but in the application he must set forth, as above, one or more of the intended methods.

4. State how long the trade-mark has been in use, and name 1 or more foreign countries with which it is used in commerce.

5. Send us twelve copies of the trade-mark.

Also remit at the same time $45 in full for the expenses, of which $25 are for Government fees and $20 Munn & Co.'s charge.

We will then prepare the necessary petition, declaration, and affidavit for signature by the applicant, and, shortly after filing the papers in the Patent Office, the official certificate will be forwarded to him.

The right to the use of any trade-mark is assignable by an instrument of writing, and such assignment, to insure its validity, should be recorded in the Patent Office within sixty days after its execution.

For assignments, searches of trade-marks, etc., address MUNN & CO., 361 Broadway, New York.

DECISIONS.—The word "star," if registered as a trade-mark, is infringed by the use of the figure of a star, and *vice versâ*.

As a rule, a geographical name can not be registered as a trade-mark, but may become so when not descriptive. For example, the words "German Syrup" held to be a lawful trade-mark.

Trade-marks such as the following may be registered: For tobacco plugs, "Andrew Jackson;" for military goods, "Smith & Co.," combined with the figure of two crossed swords; for cigars and tobacco, the letters "B. C.," no matter how arranged; for a medical compound, the words "Great American;" for the same, "Bennington's," with a portrait of Dr. Bennington; for shirtings, the figure of a peacock; for dry-goods, the words "There's millions in it;" for pickles, the words "Thunder and Lightning;" for edge-tools, the word "Washoe;" for pianoforte, the word "Weber;" for soap or cosmetic, the word "Hypatia."

ON smooth ice ôn Hudson River, velocity of wind twenty miles per hour, the best ice-boats sail sixty miles per hour, or three times faster than their wind.

COPYRIGHTS FOR LABELS.

The copyright law (see p. 99) contains a provision for the registration of copyrights on labels; but in a decision made by the Supreme Court of the United States, May 11, 1891 (O. G. 55), it was held that the copyright law has reference only to such writings and discoveries as are the result of intellectual labor. It does not have reference to labels which simply designate or describe the articles to which they are attached. To be entitled to a copyright, the article must have by itself some value as a composition, at least to the extent of serving some purpose other than as a mere advertisement or designation of the subject to which it is attached.

This decision greatly limits the issue of label copyrights. The cost for label registration is $20, of which sum $6 is the Government fee and $14 the agency fee. Address MUNN & Co., 361 Broadway, New York.

"EXPERIMENTAL SCIENCE."

The question is frequently asked, "How can I best make myself acquainted with such subjects as Electricity, Sound, Heat, Light, Photography, Microscopy, the general principles of Hydraulics, Pneumatics, Mechanics, Natural Philosophy, etc.?" In reply, we say, judging from the practical experience of others, there is no better method and no better instructor than the noble volume entitled "Experimental Science," by Geo. M. Hopkins. This is a book of 740 pages, illustrated with 680 fine engravings. Price $4.

It treats of the various topics of physics in a popular and practical way. It describes the apparatus in detail, gives dimensions, and explains the experiments in full, so that any intelligent person may readily make the apparatus, and obtain a positive experimental knowledge of each subject. The great Faraday insisted that actual experiment was the best way to acquire definite information.

The aim of the writer has been to render experimentation simple and attractive.

This book is a magazine of simple and instructive experiments, adapted to fix the knowledge in the mind of the reader. The amateur electrician will here find simple instructions in the measurement of resistance, and of electro-motive force and current strength; the best arrangement of batteries; the construction of dynamos and motors. He will also. find the principle of the dynamo, the telephone, and other electrical machines and apparatus fully explained.

As a gift nothing could be more appropriate or acceptable than a copy of this book. The illustrations cost over ten thousand dollars. The printing is done on heavy paper in the best style of the art. It is the most thoroughly illustrated work ever published on experimental physics.

Published by MUNN & Co., at the office of the SCIENTIFIC AMERICAN, 361 Broadway, New York.

COPYRIGHTS FOR BOOKS, PAMPHLETS, ETC. 33

COPYRIGHTS FOR BOOKS, PAMPHLETS, CHARTS, PICTURES, AND ART WORKS.

ANY citizen or resident of the United States may obtain a copyright who is the author, inventor, designer, or proprietor of any book, map, chart, dramatical or musical composition, engraving, cut, print, or photograph or negative thereof, or of a painting, drawing, chromo, statue, statuary, and of models and designs, intended to be perfected as works of the fine arts.

A copyright is not valid unless the title or description is recorded in the Library of Congress *before the publication of the work.*

Those who desire to obtain copyrights are requested to communicate with MUNN & CO., No. 361 Broadway, New York, and send us *the title* of the book, print, photograph, or article. We will then cause the title to be printed, and recorded at Washington, as by law required. The official certificate of copyright will then be immediately sent to our client. Our charge to attend to the business of obtaining a copyright is $5, which please remit with the title. Copyrights are filed in advance of the issue of the work; therefore we only need to receive from the applicant *the intended title* of his production, not the work itself.

If a copyright is desired for a painting, drawing, chromo, statue, statuary, or model or design for a work of art, send us the intended title and also a brief description thereof and $5.

Copyrights are granted for the term of twenty-eight years, and may be renewed for fourteen additional years, if the renewal is filed within six months before the expiration of the first term.

Copyrights may be assigned; the assignment must be recorded by the Librarian of Congress.

Foreigners who are not residents of the United States can obtain copyrights. For conditions, see terms of the law on page 93.

Labels for goods, bottles, etc., may be copyrighted. Cost, $20. See page 32. But machines and inventions cannot be copyrighted.

Address MUNN & CO., No. 361 Broadway, New York, for further information.

The intensity of illumination on a given surface is inversely as the square of its distance from the source of light. If the page of a book held twelve inches from a candle be moved six inches nearer, the light on the page is made four times stronger.

The average velocity of light is 185,000 miles per second. The light from the sun occupies 8¼ minutes in traveling to the earth, the distance being nine-two millions of miles. The light of the fixed star "Sirius," supposed to be the nearest of the stars, is 3¼ years in reaching the earth, the distance being over twenty millions of millions of miles.

MECHANICAL DRAWING.

By Prof. C. W. MacCord, of the Stevens Institute of Technology. Specially prepared for the SCIENTIFIC AMERICAN SUPPLEMENT. A series of new, original and practical lessons in mechanical drawing, accompanied by carefully prepared examples for practice, with directions, all of simple and plain character, intended to enable any person, young or old, skilled or unskilled, to acquire the art of drawing. No expensive instruments are involved. Any person with slate or pencil may rapidly learn. The series embodies the most abundant illustrations for all descriptions of drawing, and forms the most valuable treatise upon the subject ever published, as well as the cheapest. The series is illustrated by upward of 450 special engravings, and forms a large quarto book of over 100 pages. Price, stitched in paper, $2.50. Bound in handsome stiff covers, $3.50. Sent by mail to any address on receipt of price. MUNN & CO., publishers, N. Y.

For the convenience of those who do not wish to purchase the entire series at once, we would state that these valuable lessons in mechanical drawing may also be had in the separate numbers of SUPPLEMENT, at ten cents each. By ordering one or more of the following numbers at a time, the learner in drawing may supply himself with fresh instructions as fast as his practice requires. These lessons are published successively in SUPPLEMENTS Nos. 1, 3, 4, 6, 8, 9, 12, 14, 16, 18, 20, 22, 24, 26, 28, 30, 32, 36, 37, 38, 39, 40, 41, 42, 43, 44, 45, 46, 47, 48, 49, 50, 51, 52, 53, 54, 56, 58, 60, 62, 65, 69, 74, 78, 84, 91, 94, 100, 101, 103, 104, 105, 106, 107, 108, 134, 141, 174, 176, 178.

MUNN & Co., Publishers, New York.

☞ A catalogue of valuable scientific papers contained in SUPPLEMENT sent free to any address.

AIDS TO DRAWING.—Simple and plain directions how to make the Pantograph for copying drawings, also how to make the Camera Lucida for copying, also the Camera Obscura, the Reflecting Drawing Board, the Transparent Drawing Table, and other instruments, contained in SUPPLEMENT 158, price ten cents. MUNN & Co., No. 361 Broadway.

RIGHTS OF EMPLOYERS AND EMPLOYEES.

THE Supreme Court of the United States, in the case of the Union Paper Collar Company (*Official Gazette*, 1875), decides substantially as follows in respect to the rights of employers and employés, touching the proprietorship of new inventions:

Where a person has discovered a new and useful principle in a machine, manufacture, or composition of matter, he may employ other persons to assist in carrying out that principle; and if they, in the course of experiments arising from that employment, make discoveries auxiliary to the plan and preconceived design of the employer, such suggested improvements are, in general, to be regarded as the property of the party who discovered the original principle, and they may be embodied in his patent as part of his invention. Doubt upon that subject can not be entertained.

But persons employed as much as employers are entitled to their own independent inventions: and if the suggestions communicated by the persons employed constitute the whole substance of the improvement, the rule is otherwise, and the patent, if granted to the employer, is invalid, because the real invention or discovery belongs to the person who made the suggestions.

The doctrine held by the Patent Office is that an inventor who is an employer has the right to avail himself of the mechanical skill of those whom he employs to put his invention into practical form. If the inventor-employer gives general directions to his workmen to produce a certain machine, the combination or parts, or arrangement produced, belongs exclusively to the inventor-employer, and the workman has no patentable right therein.

But when a workman himself suggests and invents an improvement, without previous direction from his employer, the invention belongs to the workman; he can patent it, and the employer has no claim thereon, although the device may have been made in the shop of the employer, with his tools, and during time belonging to him.

BE neither lavish nor niggardly; of the two, avoid the latter. A mean man is universally despised, but public favor is a stepping-stone to preferment; therefore, generous feelings should be cultivated.

THE speed of an electric spark, travelling over a copper wire, has been ascertained by Wheatstone to be two hundred and eighty-eight thousand miles in a second.

36

THE SCIENTIFIC AMERICAN OFFICE, NEW YORK.

THE SCIENTIFIC AMERICAN OFFICES.

In the marvellous industrial progress of the United States during the last half century, nearly, the SCIENTIFIC AMERICAN has assisted and shared. In common with all good enterprises it has grown with the growth of the country. Many of the great industries of the present day, numbers of important manufacturing towns and districts, had no existence when our first office was opened. The date of the first number of the SCIENTIFIC AMERICAN is August 28, 1845. A back room, about twelve feet square, on an upper story in a building on Fulton St., New York, then sufficed for all our wants.

The chief offices of the SCIENTIFIC AMERICAN are now located in the splendid fire-proof building number 361 Broadway, corner of Franklin St.—named for the great philosopher and inventor. The engraving on the opposite page shows the exterior of the edifice; an interior view is given on page 38. Our building has a frontage of 58 feet on Broadway and 160 feet on Franklin St. Its location is very central and convenient—only a few steps from the Post Office, the City Hall Park, the Brooklyn Bridge, and all the converging lines of street cars. Our building is of iron, with brick arched floors, constructed in the most substantial manner, heated by steam, supplied with elevators, and the best conveniences for the rapid transaction of business. The visitor enters a handsome elevator car at the entrance door on Broadway, and in a twinkling is landed directly within our office. Our main office occupies an entire floor of the building, and is a grand apartment, flooded with light. Here are the subscription and advertising departments of the SCIENTIFIC AMERICAN, and here is prosecuted the principal part of our labors in connection with the preparation of patent drawings and specifications pertaining to the SCIENTIFIC AMERICAN Patent Agency.

In addition to the transaction of an immense amount of patent business, which comes to us by mail and express, large numbers of authors and inventors prefer to come in person to our offices to explain the merits of their improvements.

In carrying on our extensive patent business, we aim to conduct it in the most expeditious and systematic manner. We are assisted by the most experienced examiners and specification writers. The finest mechanical draughtsmen in the country prepare our drawings.

On the next floor above our main office is our model-

INTERIOR OF THE SCIENTIFIC AMERICAN OFFICE.

THE SCIENTIFIC AMERICAN OFFICES. 39

room. All models are ticketed with the names of their owners, and here carefully stored; no one, except our confidential assistants in charge, being permitted to enter. All correspondence relating to patents is carefully preserved for a given time in large fire-proof safes of extra thickness.

All drawings of pending patent cases are likewise preserved in similar safes, specially made for us for this purpose.

The utmost care is taken to guard the privacy and preserve the safety of the many thousands of inchoate inventions committed to our care; and we may here mention with satisfaction the fact that during our long professional career of over thirty years, not one of our clients has ever found his confidence in us misplaced.

The remaining divisions of our main establishment in New York are the type-room and the editorial-rooms where the interesting matter that fills the pages of THE SCIENTIFIC AMERICAN is prepared.

THE MODEL ROOM.

The printing of THE SCIENTIFIC AMERICAN is done in a neighboring building, where several large steam presses are kept in constant motion, day and night, during the greater part of each week, to work off our large edition. After leaving the press the sheets pass through a folding-machine, are by machinery then trimmed, and then enveloped for the mail.

The addresses of our subscribers are printed on slips of paper, and, during the mailing operation, cut and attached by means of a curious little instrument to the separate copies. The regular combined edition of THE SCIENTIFIC AMERICAN and SCIENTIFIC AMERICAN SUPPLEMENT at present is very large. All back numbers of the SUPPLEMENT are kept on hand, 10 cents each. Yearly subscription, $5 a year. Regular subscribers receive the paper free of postage, which is paid by us. The terms of subscription to THE SCIENTIFIC AMERICAN are $3.00 a year. It is sold by single numbers by all news dealers, 10 cents per number. Specimen copies can also be had by sending 10 cents to MUNN & CO., 361 Broadway, New York.

OUR BRANCH OFFICE IN WASHINGTON.

WE have a large office in Washington, located at 622 F. street, Pacific Building, very near Seventh street and the Patent Office,'

The first building on the right is the United States Patent Office; the next on the same side is the General Post Office; on the left stands THE SCIENTIFIC AMERICAN office. Our location is especially convenient for the transaction of business. We employ at Washington a corps of trained assistants, part of whom make it their exclusive duty to watch and assist the progress of our cases before the Patent Office. For these services we make no extra charges.

Another division of our helpers in Washington devote themselves to the preliminary examination of inventions—a matter that is explained more fully on page 17.

It is to the systematic method and abundant supply of trained helpers, personally supervised by the proprietors, that the long-continued prosperity and remarkable success of THE SCIENTIFIC AMERICAN Patent Agency are due.

NO TAXES ON PATENTS.

THE patent and all its rights are under the owner's control; and after a patent is issued it is not subject to additional payments or to taxes of any kind, whether national, State, or local.

WILL IT PAY?

ON page 16 readers are informed that we are always happy to give them our opinion as to the novelty of their inventions, *without charge*. But some persons, when they send for such information, add many other inquiries, difficult to answer, and not included in our gratuitous invitation; as for example: "What is it worth? Who will buy? Will it pay? Does it infringe? Does it conflict with B's patent? If you will guarantee it does not infringe, I will apply for a patent," etc.

The following hints may prove useful as a sort of general answer.

"What is it worth? Who will buy?" As a general ruie, an invention is worth little or nothing until the patent is obtained; and until then no one is likely to buy. Therefore the first thing to be considered, the first step to be taken, is to *obtain the patent*.

SOUND.

"Will it pay?" As a general rule, every patentable improvement will more than repay the small cost of taking out the patent. The sale of a single machine, or of a single right of use, will often bring back more than the whole outlay for the patent. The extent of profit frequently depends upon the business capacity of the inventor or his agent. One man by his activity will make a fortune from an unpromising improvement, while another, possessing a brilliant invention, will realize little or nothing, owing to incompetence.

SOUND

Is the effect produced upon the ear when air is set in motion within certain limits of rapidity. Audible sound begins when about thirty-two vibrations per second are made, and ceases when about 40,000 vibrations per second are reached. In an organ, the deepest note has thirty-two vibrations per second, the highest, 3480. The compass of the human voice is, on an average, about two octaves. Deep F of a bass singer has 87 vibrations per second; upper G of treble, 775.

The number of vibrations corresponding with the middle C of a musical instrument is 522 per second. An octave below, half the number; an octave above, twice the number.

Sound travels at the rate of 1100 feet per second in a still atmosphere. The distance in feet between an observer and the point where a stroke of lightning falls, may be known by multiplying 1100 by the number of seconds that elapse after the flash is seen until the sound is heard.

MELTED snow produces from $\frac{1}{8}$ to $\frac{1}{4}$ of its bulk in water.

OCEAN waves rise from 20 to 22 feet in extreme height, at which altitude there are 3 in a mile and 4 per minute.

THE highest heat of a common wood fire is estimated at 1140° F.

NEARLY all solids become luminous at 800 degrees of heat F.

THE force of expansion of solids by heat is enormous. Thus iron, if heated from 32° F. to 212°, expands .0012 of its length, to produce which change of length by mechanical means would require a force of 15 tons.

THE best engines and boilers develop a horse-power per hour by the consumption of two pounds of coal. But this is better than the average; and three pounds of coal per horse-power, per hour, is a more common result.

MODEL-HALL, UNITED STATES PATENT OFFICE, WASHINGTON.

42

THE PATENT OFFICE AT WASHINGTON.

The situation of the SCIENTIFIC AMERICAN Patent Agency in Washington, on F Street, near Seventh Street, is so remarkably convenient that we are enabled to transact a large amount of patent business very quickly. As all models, drawings, specifications, trade-marks, records of assignments, etc., are deposited in the Patent Office, we have access thereto, on behalf of our clients, by simply stepping across the street.

Next to the Patent Office, on the right, directly opposite THE SCIENTIFIC AMERICAN, is the General Post Office. Here the Postmaster-General sits, and the postal service of the country is regulated. The Post Office building is 304 feet long and 204 feet wide, Corinthian style, of white marble.

We present two engravings illustrative of the Patent Office.

The engraving on page 24 shows a full exterior view of the Patent Office, which is one of the finest edifices in Washington. It is of the Doric order of architecture, 433 feet long, 331 feet wide, 75 feet high. The collection of models of inventions here gathered is very remarkable. the aggregate number being over two hundred thousand. Nearly twenty thousand new models are sent to the Patent Office each year.

On page 42 we give an interior view of one of the great model-rooms of the Patent Office, nearly 400 feet long, paved with marble. The models, it will be observed, are deposited in glass cabinets upon the main floor and galleries. The model-rooms are open to the public from 9 A. M. to 3 P. M.

No printed statement or recommendation that we could present will convey to the mind of the visitor so adequate and truthful an impression of the magnitude and wonderful success of our (Munn & Co.'s) labors in procuring patents for inventors as a walk through the Patent Office. The visitor beholds tier upon tier of models, rising on both sides from floor to ceiling, occupying a main portion of the entire building, and finds, on examining the records, that every cabinet, every class of invention, is crowded with models sent from THE SCIENTIFIC AMERICAN Patent Agency, and that a very large portion of all the patents granted are to our (Munn & Co.'s) clients.

The number of persons generally employed at the Patent Office is between four and five hundred. The principal officers are the Commissioner of Patents, who is the executive, the Assistant Commissioner, and seventy examiners.

Their aggregate salaries amount to about one hundred and fifty thousand dollars a year. One hundred and thirty thousand dollars a year, nearly, are paid into the Patent Office by THE SCIENTIFIC AMERICAN Patent Agency alone.

THE LARGEST AND BEST.

Now and then professional rivals, jealous or unreasonable persons, will be found who rail about Munn & Co., usually because we transact so much business and make small charges. These carpers falsely allege that they can do the business better, and afford more time.

Everybody knows, however, that the best service and the most reasonable rates are generally furnished by large, well-conducted establishments, and the patent-agency business is no exception. We have a staff of trained assistants and draughtsmen; we give to every case careful study, experienced care, and *abundance of time;* we have helpers at Washington who make it their special duty to watch over and assist the progress of our cases before the Patent Office, give explanations, and see that the best claims are allowed. No extra charges are made for these services. Our efforts are usually successful and give general satisfaction. For many years we have secured more patents for inventors and done more patent business than the combined business of the majority of the four hundred patent agents in this country. A very large proportion of all the most valuable and successful patents now existing were obtained through THE SCIENTIFIC AMERICAN Patent Agency.

FOREIGN PATENTS.

THE American patent law contains a special provision for the benefit of the inventor in respect to foreign patents, namely: It provides that after a home patent is allowed, the invention may remain in the secret archives of the Government for a period not exceeding *six months*, if the applicant so desires, thus enabling him to arrange for patents in foreign countries in advance of all other persons.

If the inventor is unable to meet the expenses of the foreign patents, he should find a reliable assistant or partner who will pay the costs and share the profits. Partnerships of this kind have in many cases proved highly profitable to all concerned. Arrangements with one partner for England, another for France, and so on, are suggested.

It is of the utmost importance to the interests of the applicant in taking out foreign patents that he should employ home agents for their procurement, who are well known for responsibility, experience, and integrity. He is thus enabled to obtain better service and a better patent, together with prompt information as to the condition of his patent and the steps necessary for its preservation.

CANADA.

THE expense to apply for a Canadian patent is forty dollars ($40), which includes Government tax, agency, and all charges for five years, after which two additional terms of five years each may be obtained on payment of twenty-five dollars each —in all fifteen years. The patent may be applied for at the outset for fifteen years, at a cost of eighty dollars ($80). Inventions that have been already patented in the United States for not more than one year may be secured in Canada by the inventor, who must sign the papers. If patented for more than one year in the United States, the Canadian patent is refused.

In order to apply for a patent in Canada, please send to us (MUNN & CO.) a description and drawing of the invention, and remit $40. If already patented in the United States, a copy of the patent should be supplied. We will then immediately prepare the documents and forward them to the applicant for his signature. His personal presence is unnecessary. All the business can be promptly done by correspondence. The time required to secure the patent is from four to six weeks; the patent is granted without a model, but before the document is actually delivered a small model must be furnished.

During the first year of a Canadian patent the holder may import the patented article ready made. Within two years from the date of the patent he must begin the manufacture in Canada, or arrange for some place where persons wishing the invention can order the same. The Canadian patent covers Nova Scotia, Prince Edward Island, and both the Canadas.

GREAT BRITAIN.

THE British patent extends over England, Wales, Scotland, Ireland, and the Channel Islands, but not the Colonies; the latter make their own patent laws.

Under the new law, the expense for an English patent is fifty dollars ($50), which includes Government taxes, agency, and all charges for the first period or provisional patent, good for nine months. A second instalment of $50 is payable in New York eight months after the date of the provisional patent. The patent is then completed and the great seal attached. It is generally better to take the complete patent at first, which costs $100, and takes the place of the two payments just mentioned. No further taxes are payable until the end of the fourth year, when $50 are due; after that, annual payments, slightly increasing, are required each year. The patent is granted for fourteen years, but ceases if any tax is not duly paid. The patentee in Great Britain possesses the same full and exclusive rights as in the United States, and the patent may be assigned in the usual manner.

Great Britain has a population of *forty millions*, and is one of the principal financial, commercial, and manufacturing centres of the world. The importance to our citizens of securing English patents for their new inventions can not be overrated.

FRANCE AND BELGIUM.

THE cost to apply for a French patent is one hundred dollars ($100), which covers agency fee, tax, and all expenses for the first year. No official examination is made; no model. The term of the patent is fifteen years, subject to annual tax of $20. The patent ceases when any due tax is unpaid.

Belgium is the manufacturing centre for a large portion of the Continent, and Belgian patents rank among the most desirable of those that are taken out by American citizens.

The expense to apply for a Belgian patent is one hundred dollars ($100). The law and proceedings are substantially the same as in France. A small tax is payable annually. The longest term of the patent is twenty years.

GERMANY AND OTHER COUNTRIES.

IN the following countries the cost of applying for the patent varies with the period of the grant, which may generally be from five to fifteen years, at the option of the applicant. The costs to apply for a patent for the shortest term are: in Germany, $100; Austria, $100; Norway, $100; Sweden, $100; Denmark, $100; Spain, $100; Italy, $150; Russia, $300; Portugal, $400; Brazil, $000. No patents granted in Switzerland and the Netherlands.

Patents are also granted in the British Colonies and the several South American States.

N.B.—We would remind all who desire to take Foreign Patents, that (1) we have the best facilities for work; (2) we have had over thirty-seven years' experience; (3) the Foreign Patent, when issued, is *noticed without charge* in the SCIEN-

HINTS ON THE SALE OF PATENTS. 47

TIFIC AMERICAN, which has a large circulation in Europe. This publication is often copied into other papers, and invariably assists the introduction and sale of the patent.

We furnish, *free*, a pamphlet containing additional particulars, and shall be happy to give any other information that may be desired. Address MUNN & CO., Solicitors of American and Foreign Patents, 361 Broadway, New York.

HINTS ON THE SALE OF PATENTS.

THE original study and planning of a new thing is usually a labor of love on the part of the author. The work is suitable for the leisure hour, the winter's evening, the quietude of home. The plan being finished, then comes the business of introduction and sale. The first step in the material progress of the invention is its development into the form of a *public record* on which the patent issues. This business requires experienced skill for its proper transaction, and the inventor will generally promote his own interests by employing trustworthy solicitors. Not so, however, in respect to the second step, namely, the *making of money* out of the patent. This is a commercial proceeding, involving the ordinary details of industry, prudence, and care. The patentee himself is generally the best manager in this department.

The first thing to be done after receiving a patent is to *make known the merits of the invention as widely as possible*. This is like ploughing and seeding the ground. If well done, the crop will grow, even while the husbandman sleeps.

One of the quickest and most effective methods of bringing the merits of an invention before the public, is to have it noticed and engraved in THE SCIENTIFIC AMERICAN. This paper, published weekly, is seen by probably not less than *three hundred thousand readers*, who comprise all of the most intelligent persons of scientific and mechanical acquirements in the country. The fact of publication in THE SCIENTIFIC AMERICAN is a passport to their attention and favor. "Yes, that is a good invention. I have seen it illustrated in THE SCIENTIFIC AMERICAN, and understand its construction. I advise you to purchase the right." We suppose that more patents are sold upon such advice than by all other agencies and means put together.

The splendid engravings which adorn our paper are prepared by the most talented artists. We are always glad to illustrate new and useful inventions in THE SCIENTIFIC AMERICAN, and, owing to the interest which our readers take in such novelties, we make the expense to the patentee as low

48 HINTS ON THE SALE OF PATENTS.

as possible—generally but very little above the actual cost to us. If any one will take the trouble to count the probable cost to him of printing and circulating, by mail, a mass of circulars containing an engraving and description of his invention, and then compare that cost with the insignificant figure he would have to pay us to get up the same cut and description, and *print and circulate fifty thousand copies thereof in* THE SCIENTIFIC AMERICAN, he will appreciate the marvellous economy offered by our journal. The circular plan would cost the patentee more for the *white paper alone*, than we (Munn & Co.) should charge for the entire job. After being electrotyped and published, the original blocks are sent to the owner, who can then use them for other papers, circulars, letter-heads, bill-heads, etc.

Let us here remind the inventor that the value of property in patents is now far greater than in former years, when the population was sparse, and the demand for new manufactures small. Therefore do not part with your patent unless you can realize from it adequately. Any foolish person can give or throw away his property; and we are sorry to say that thousands of valuable patent privileges are wasted by their owners for lack of a little patience.

In general, the best way to begin is to manufacture the article, and also to grant licenses under the patent; unless handsomely paid, avoid the sale of any undivided interest in the proprietorship of the patent, such as a sixteenth, an eighth, or a quarter of the patent. By even one such sale, no matter how small, the patentee loses the control of his patent; under the license plan he does not.

It must not be supposed, because a patent is granted, that the world will run after an unknown man to buy from him an unknown patent. In order to sell licenses or rights under a patent, judicious effort is required on the part of the inventor. Indeed, his final success will depend, to a considerable extent, upon his business tact and energy. He should make himself thoroughly conversant with the merits of his invention, and should prepare specimens or model machines thereof, made in the most perfect manner, so as readily to exhibit the operations of the improvement to others.

A very profitable method of realizing from a patent is to grant town or county licenses, employing good and reliable special agents to travel about and sell them. Such agents expect to make money by the operation, and generally need to have a liberal allowance of the proceeds devoted to their remuneration. In the example of a bee-hive patent, the patentee might issue to the agent, duly signed, a number of county licenses, not good, however, until countersigned by the

HINTS ON THE SALE OF PATENTS. 49

agent. Suppose the price for a county is fixed at ten dollars per thousand inhabitants. The agent deposits with the patentee twenty-five dollars or other agreed sum on each license, to be returned if he fails to sell. He, however, sells a county containing 20,000 population for $200, retains by agreement half the proceeds, or $100, and returns $100 to the patentee. The foregoing will be suggestive of many other methods of disposing of patents by special agents, which is usually the most lucrative method of procedure.

In some cases an excellent method is to commence the manufacture of the article in a suitable locality, and when it is so far under way as to exhibit progress and merit, then to sell out the business with license under the patent. This method is often very remunerative.

The patentee may subdivide his patent into as many different classes of rights as he chooses, and sell each class by separate agents or otherwise, as he prefers. Thus, the patentee of a sewing-machine may license one party to sew straw goods, another party to sew cotton goods, another silk, another woollen, etc.

The patentee may, if he desires, require purchasers of his machines to pay him a regular annual rental for the use of the machine, or a tariff upon the goods produced, in addition to the original price of the machine. Thus in the case of the wood-planing machine the patentees required, say, $5000 to be paid in cash, for which they allowed the licensee to build one machine; and thereafter, for every foot of lumber planed by the machine, the patentee received an additional payment or royalty.

In many of the States general laws exist for the incorporation of manufacturing companies for the development of improved articles. The usual method, where a patent is to form the basis of such a company, is for several persons—three or more—to unite under some agreed title, appoint trustees, president, treasurer, secretary, fixing capital stock at any desired sum, say fifty thousand dollars. The incorporators contribute in money say twenty-five thousand dollars, and take half of the stock; and they issue twenty-five thousand dollars in fully-paid-up stock to the patentee, who assigns to the company the agreed right or license under the patent. These companies, if properly managed, are often highly profitable. The patentee should see to it that the required amount of cash capital is actually paid up into the treasury before delivering his assignment, thus insuring the effective working of the invention. Several distinct companies may be organized in this manner, in different places, on the basis of one good patent.

PROFESSIONAL PATENT-SELLERS.

The license and royalty plan is often a most profitable method of employing patents. This, in effect, involves a sort of contract between a patentee and a partner or manufacturer, by which the latter, in consideration of license to make the thing, agrees to pay to the patentee a specified sum upon each article made or sold. The patentee of the chimney-spring, now so commonly used to fasten glass chimneys upon lamps, was accustomed to grant licenses to manufacturers on receiving a royalty of a few cents per dozen. His income was at one time reported to be fifty thousand dollars a year from this source. Howe, the inventor of the sewing-machine, received a royalty of from five to ten dollars on each machine, and his annual income was estimated at five hundred thousand dollars. Goodyear, the inventor of vulcanized rubber, divided his patent up into many different rights, licensing one company for manufacturing rubber combs, another for hose-pipes, another for shoes, another for clothing, another for wringers, etc. Each company or partner paid a tariff. Lyall, inventor of the continuous loom, has in like manner divided his patent into many different rights: one company weaves carpets, another corsets, another bags, another sheetings, and so on. He enjoys an enormous income from his invention. We might give many similar examples.

Licenses, shop rights, rights of use, if not exclusive, need not be recorded at Washington. But a grant of an exclusive territorial right, or of an undivided interest in a patent, should be recorded. The business may be quickly done through Messrs. Munn & Co. See page 27.

PROFESSIONAL PATENT-SELLERS.

No sooner does any person's name appear in print as the patentee of a new invention, than he receives, by mail, a shower of letters and circulars, from individuals who set forth that they have remarkable facilities for the selling of patents. The patentee is invited, if he wants to realize immediately, say one thousand, two thousand, or ten thousand dollars, to signify his desire to that effect, and send forward to the agent *a small advance fee*. Thus, instead of helping the patentee to obtain money, they begin by drawing money from him; upon this they live and flourish. We are often asked if these people, who so pressingly and plausibly claim to be able to sell patents, are reliable, and whether they ever effect sales. We regret to be obliged to say that we seldom or never hear of their making any sales. There are twenty thousand new applicants for patents every year, from whom these pretending sellers obtain money. They busy themselves in writing letters to inventors and in working them up to the remitting point, but have no time left for the drudgery of patent-selling even if they had any ability in that direction. There is no trickery too low for some of these sellers: one concern, for example, has gone so far as to imitate and adopt our long-established firm name of Munn & Co. But we do not sell patents, nor have we connection with any concern that pretends so to do. The truth is, that the profit upon the sales of a single good patent is equivalent to a fortune, and the business it furnishes is enough to fully engage the attention of many persons. Our advice to patentees is: Take hold yourselves of the business of selling. If you want assistance, search for suitable agents among your friends, and interest them specially in your invention.

HOLD THE FORT.

IF you have made an invention for which you desire to secure a patent, but lack the necessary funds, do not for that reason be so foolish as to give or to throw away the discovery ; do not part with any considerable portion for a pittance ; do not, as is so commonly the case, promise or convey a half or any undivided portion of the improvement. If you are pinched for money you can generally, by patience and perseverance, obtain the use of the small sum required, by explaining the merits of the invention to intelligent, reliable persons in your vicinity. To the party who is disposed to make the desired loan, the grant of a privilege for a town or county will generally be a satisfactory recompense, especially if he believes that it will really assist you in the further development of your invention. The following conveyance will, in general, be ample in such cases :

"Whereas I, Richard Roe, of Wyoming, County of Mohawk, State of New York, have invented a new and useful improvement in musical instruments, for which I am about to apply for letters-patent; and whereas John Doe, of Wyoming, New York, hath advanced to me the sum of one hundred dollars towards the expenses of said patent:

"Now this indenture witnesseth, that for and in consideration of said payment to me made, I do hereby grant and convey to the said John Doe, his heirs or assigns, a license to make, use, and sell the invention, within the limits of the County of Mohawk, State of New York, for and during the full end of the term for which said letters-patent are or may be granted. Witness my hand and seal, this first day of January, A.D. 1892.

"In presence of   RICHARD ROE.
"W. LOE."

STATE LAWS CONCERNING PATENT-RIGHTS.

IN some of the States, laws have been passed by which patentees or their agents who offer *patent-rights* for sale, without complying with certain State regulations, are made liable to fine and imprisonment.

The United States Court, in the case of John Robinson; held that this kind of legislation is unauthorized, that property in inventions exists by virtue of the laws of Congress, and that no State has a right to interfere with its enjoyment, or annex conditions to the grant. If the patentee complies with the laws of Congress on the subject, he has a right to go into the open market anywhere within the United States and sell his property. If this were not so, a State might nullify the laws of Congress and destroy the powers conferred by the Constitution.

All laws of State legislatures that in any manner interfere with the free sale of patent-rights, such as the requiring of the agent or patentee to file copies of patent, take licenses, procure certificates, comply with forms, or which release the payee of ordinary notes of hand given for patents, have been declared unconstitutional and void by the United States Courts.

All State judges, sheriffs, or other State officials who undertake to interfere with patentees or their agents in the free sale of patents, make themselves liable in damages and other punishment.

The decisions of the United States Courts on these points are given in SCIENTIFIC AMERICAN SUPPLEMENT, No. 25. Price 10 cents. To be had at SCIENTIFIC AMERICAN Office, 361 Broadway, and at any news store.

The above decisions apply only to the sale of *patents and rights under patents*, not to the peddling of goods or the sale of manufactured articles. All citizens of the United States must comply with the usual local license laws concerning the sale of goods, whether the goods are patented or not. But no State can lawfully enact a special law adverse to the sale of patented goods, or impose any special restriction, tax, or fine upon persons who go about to sell patented goods or patented articles of any description.

PLATINUM has been drawn into wires only one thirty-thousandth ($\frac{1}{30000}$) part of an inch, invisible to the eye, and one mile's length weighing only one grain.

A CUBIC foot of air weighs 535 grains. Water is 815 times heavier than air. A cubic foot of water weighs 62½ lbs., a gallon 8 4/10 lbs.

ABSTRACT

FROM THE

RULES OF PRACTICE

IN THE

UNITED STATES PATENT OFFICE.

WHO MAY OBTAIN A PATENT.

ANY person, whether citizen or alien, being the original and first inventor or discoverer of any new and useful art, machine, manufacture, or composition of matter, or any new and useful improvement thereof, may obtain a patent for his invention or discovery, subject to the conditions hereinafter named.

In case of the death of the inventor, the patent may be applied for by, and will issue to, his executor or administrator. In case of an assignment of the whole interest in the invention, or of the whole interest in the patent if granted, the patent will issue to the assignee, upon the request of the latter, or his assignor; and so, if the assignee holds an undivided part interest, the patent will, upon a similar request, issue jointly to him and the inventor; but the assignment must first have been entered of record, and at a day not later than the date of the payment of the final fee. The application and oath must be made by the actual inventor, if alive, even if the patent is to issue to an assignee; but where the inventor is dead, the application and oath must be made by his executor or administrator.

Joint inventors are entitled to a joint patent; neither can claim one separately; but the independent inventors of separate and independent improvements in the same machine can not obtain a joint patent for their separate inventions; nor does the fact that one man furnishes the capital and the other makes the invention entitle them to make application as joint inventors.

54 RULES OF PRACTICE.

A patent will not be granted to an applicant if what he claims as new has been, before his invention, patented or described in any printed publication in this or any foreign country, or been invented or discovered in this country, nor if he has once abandoned his invention, nor if it has been in public use or on sale more than two years previous to his application.

If it appears that the inventor, at the time of making his application, believes himself to be the first inventor or discoverer, a patent will not be refused on account of the invention or discovery, or any part thereof, having been known or used in any foreign country before his invention or discovery thereof, it not appearing that the same, or any substantial part thereof, had before been patented or described in any printed publication.

Merely conceiving the idea of an improvement or machine is not an "invention" or "discovery." The invention must have been reduced to a practical form, either by the construction of the machine itself or by such disclosure of its exact character that a mechanic, or one skilled in the art to which it relates, can and does construct the improvement before it will prevent a subsequent inventor from obtaining a patent.

No application for a patent will be regarded as complete, or be placed upon the files for examination, until the fee is paid, the specification, the petition, and the oath, properly signed, are filed, and the drawings and a model or specimens (when required) are furnished.

Two or more separate and independent inventions can not be claimed in one application · but where several distinct inventions are dependent upon each other and mutually contribute to produce the new result, they may be so claimed.

DRAWINGS.

The applicant for a patent is required by law to furnish a drawing of his invention, where the nature of the case admits of it. The following rules will therefore be rigidly enforced, and any departure from them will be certain to cause delay in the examination of an application for letters-patent:

a. Drawings should be made upon paper stiff enough to stand in the portfolios, the surface of which must be calendered and smooth. Indian ink, of good quality, to the exclusion of all other kinds of ink or color, must be employed, to secure perfectly black and solid work.

b. The size of a sheet on which a drawing is made should be exactly 10 by 15 inches. One inch from its edges a single mar-

RULES OF PRACTICE. 55

ginal line is to be drawn, leaving the "sight" precisely 8 by 13 inches. Within this margin all work and signatures must be included. One of the smaller sides of the sheet is regarded as its top, and, measuring downward from the marginal line, a space of not less than 1¼ inches is to be left blank for the insertion of title, name, number, and date.

c. All drawings must be made with the pen only, using the blackest Indian ink. Every line and letter (signatures included) must be *absolutely black.* This direction applies to all lines, however fine, to shading, and to lines representing cut surfaces in sectional views. All lines must be clean, sharp, and solid, and they must not be too fine or crowded. Surface shading, when used, should be left very open. Sectional shading should be by oblique parallel lines, which may be about one twentieth of an inch apart.

d. Drawings should be made with the fewest lines possible consistent with clearness. By observing this rule the effectiveness of the work after reduction will be much increased. Shading (except on sectional views) should be used only on convex and concave surfaces, where it should be used sparingly, and may even there be dispensed with if the drawing is otherwise well executed. The plane upon which a sectional view is taken should be indicated on the general view by a broken or dotted line. Heavy lines on the shade sides of objects should be used, except where they tend to thicken the work and obscure letters of reference. The light is always supposed to come from the upper left-hand corner, at an angle of forty-five degrees. Imitations of wood or surface-graining should not be attempted.

e. The scale to which a drawing is made ought to be large enough to show the mechanism without crowding, and two or more sheets should be used if one does not give sufficient room to accomplish this end; but the number of sheets must never be increased unless it is absolutely necessary. It often happens that an invention, although constituting but a small part of a machine, has yet to be represented in connection with other and much larger parts. In such cases a general view on a small scale is recommended, with one or more of the invention itself on a much larger scale.

f. Letters of reference must be well and carefully formed; they are of the first importance. When at all possible, no letter of reference should measure less than one eighth of an inch in height, that it may bear reduction to one twenty-fourth of an inch, and they may be much larger when there is sufficient room.

Reference letters must be so placed in the close and complex parts of drawings as not to interfere with a thorough comprehension of the same, and to this end should rarely cross or mingle with the lines. When necessarily grouped around a certain part, they should be placed at a little distance, where there is available space, and connected by short broken lines with the parts to which they refer. They must never appear upon shaded surfaces, and, when it is difficult to avoid this, a blank space must be left in the shading where the letter occurs, so that it shall appear perfectly distinct and separate from the work. If the same part of an invention appears in more than one figure, it should always be represented by the same letter.

The foregoing rules relating to drawings will be rigidly enforced; and all drawing not artistically executed in conformity therewith will be returned to the respective applicants, or, at the applicant's option and cost, the Office will make the necessary corrections.

All reissue applications must be accompanied by new drawings, as in original applications, and the inventor's name must appear in all cases upon the same.

MODELS.

THE Patent Office does not now require a model except in special cases when called for by the examiner. It must clearly exhibit every feature of the machine which forms the subject of a claim of invention, but should not include other matter than that covered by the actual invention or improvement, unless it is necessary to the exhibition of a working model. The model must be neatly and substantially made of durable material, metal being deemed preferable; and should not in any case be more than one foot in length, width, or height. If made of pine or other soft wood, it should be painted, stained, or varnished. Glue must not be used, but the parts should be so connected as to resist the action of heat or moisture. When the invention is a composition of matter, a specimen of the composition, properly marked, must accompany the application. Specimens of the separate ingredients, if ordinary and well known, need not be furnished, unless the Office disputes their operation in the manner as stated by applicant.

THE OFFICIAL EXAMINATION.

All cases in the Patent Office are classified and taken up for examination in regular order; those in the same class

RULES OF PRACTICE. 57

being examined and disposed of, as far as practicable, in the order in which the respective applications are completed. When, however, the invention is deemed of peculiar importance to some branch of the public service, and when, for that reason, the head of some Department of the Government specially requests immediate action, the case will be taken up out of its order. These, with applications for extensions, for reissue, and for letters-patent for inventions for which a foreign patent has already been obtained, which cases have precedence over all others, are the only exceptions to the above rule in relation to the order of examination.

INTERFERENCES.

An "interference" is a proceeding instituted for the purpose of determining the question of *priority of invention* between two or more parties claiming the same patentable subject-matter. It may also be resorted to for the purpose of procuring evidence relating to the alleged abandonment or the public use of an invention.

An interference will be declared in the following cases:

First. When two or more parties have applications pending before the Office at the same time, and their respective claims conflict in whole or in part.

Second. When two or more applications are pending at the same time, in each of which a like patentable invention is shown or described, and claimed in one though not specifically claimed in all of them.

Third. When an applicant, having been rejected upon an unexpired patent, claims to have made the invention before the patentee.

The fact that one of the parties has already obtained a patent will not prevent an interference; for, although the Commissioner has no power to cancel a patent already issued, he may, if he finds that another person was the prior inventor, give him a patent also, and thus place both parties on an equal footing before the courts and the public.

REISSUES.

A reissue is granted to the original patentee, his legal representatives, or the assignees of the entire interest, when, by reason of a defective or insufficient specification, or by reason of the patentee claiming as his invention or discovery more than he had a right to claim as new, the original patent is inoperative or invalid, provided the error has arisen from inadvertence, accident, or mistake, and without any fraudulent or deceptive intention. In the cases of patents issued and assigned prior to July 8, 1870, the application for reissue may be made by the assignee; but, in the case of patents issued or assigned since that date, the application must be made and the specification sworn to by the inventor, if he be living.

The general rule is, that whatever is really embraced in the original invention, and so described or shown that it might have been embraced in the original patent, may be the subject of a reissue; but no new matter shall be introduced into the specification, nor shall the model or drawings be amended except each by the other; but, when there is neither model nor drawing, amendments may be made upon proof satisfactory to the Commissioner that such new matter or amendment was a part of the original invention, and was omitted from the specification by inadvertence, accident, or mistake, as aforesaid.

Reissued patents expire at the end of the term for which the original patents were granted. For this reason applications for reissue will take precedence, in examination, of original applications.

A patentee in reissuing may, at his option, have a separate patent for each distinct and separate part of the invention comprehended in his original patent, by paying the required fee in each case, and complying with the other requirements of the law, as in original applications. Each division of a reissue constitutes the subject of a separate specification descriptive of the part or parts of the invention claimed in such division; and the drawing may represent only such part or parts. All the divisions of a reissue will issue simultaneously. If there be controversy as to one, the others will be withheld from issue until the controversy is ended.

In all cases of applications for reissues, the original claim, if reproduced in the amended specification, is subject to re-examination, and may be revised and restricted in the same manner as in original applications. The application for a reissue must be accompanied by a surrender of the original patent, or, if lost, then by an affidavit to that effect and a certified copy of the patent; but if any reissue be refused, the original patent will, upon request, be returned to the applicant.

DESIGN PATENTS.

When the design can be sufficiently represented by drawings or photographs a model will not be required.

Whenever a photograph or an engraving is employed to illustrate the design, it must be mounted upon a thick Bristol-board or drawing-paper, ten by fifteen inches in size; and the applicant will be required to furnish ten extra copies of such photograph or engraving (not mounted), of a size not exceeding seven and a half inches by eleven.

Whenever the design is represented by a drawing, each of the ten copies must be made to conform as nearly as possible to the rules laid down for drawings of mechanical inventions.

ASSIGNMENTS.

A patent or trade-mark may be assigned, either as to the whole interest or any undivided part thereof, by an instrument of writing. No particular form of words is necessary to constitute a valid assignment, nor need the instrument necessarily be sealed, witnessed, or acknowledged.

In every case where it is desired that the patent shall issue to an assignee, the assignment must be recorded in the Patent Office at a date not later than the day on which the final fee is paid.

A patentee may not only assign the whole or an undivided interest in his patent, but he may grant and convey an exclusive right under his patent to the whole or any specified portion of the United States by an instrument in writing.

Every assignment or grant of an exclusive territorial right, as well as of an interest in a patent or trade-mark, must be recorded in the Patent Office; if a patent, within three months, if a trade-mark, within sixty days, from the execution thereof; otherwise it will be void as against any subsequent purchaser or mortgagee for a valuable consideration, without notice.

The patentee may convey separate rights under his patent to make or to use or to sell his invention, or he may convey territorial or shop rights which are not exclusive. Such conveyances are mere licenses, and need not be recorded.

The receipt of assignments is not generally acknowledged by the Patent Office; they will be recorded in their turn within a few days after their reception, and then transmitted to the person entitled to them.

The Patent Office can not respond to inquiries as to the novelty of an alleged invention in advance of an application for a patent, nor to inquiries founded upon brief and imperfect descriptions, propounded with a view of ascertaining whether such alleged improvements have been patented, and if so, to whom; nor can it act as an expounder of the patent law, nor as counsellor for individuals, except as to questions arising within the Office.

ATTORNEYS.

Any person of intelligence and good moral character may appear as the agent or the attorney in fact of an applicant, upon filing a proper power of attorney. As the value of patents depends largely upon the careful preparation of the specification and claims, the assistance of competent counsel will, in most cases, be of advantage to the applicant, but the value of their services will be proportioned to their skill and honesty. So many persons have entered this profession of late years without experience that too much care can not be exercised in the selection of a competent man. The Patent Office can not assume responsibility for the acts of attorneys, nor can it assist applicants in making a selection. It will, however, be a safe rule to distrust those who boast of the possession of special and peculiar facilities in the Office for procuring patents in a shorter time or with more extended claims than others.

FORMS.

FORMS OF PETITIONS FOR PATENTS.

By a Sole Inventor.

To the Commissioner of Patents:

Your petitioner, a resident of ———, ———, prays that letters-patent be granted to him for the invention set forth in the annexed specification.

A. B.

By Joint Inventors.

To the Commissioner of Patents:

Your petitioners, residing respectively in ———, ———, and ———, ———, pray that letters-patent may be granted to them, as joint inventors, for the invention set forth in the annexed specification.

A. B.
C. D.

By an Inventor for Himself and an Assignee.

To the Commissioner of Patents:

Your petitioner, a resident of ———, prays that letters-patent may be granted to himself and C. D., of ———, as his assignee, for the invention set forth in the annexed specification, the assignment to the said C. D. having been duly recorded in the Patent Office, in liber —, page —.   A. B.

For a Reissue, (by the Inventor.)

To the Commissioner of Patents:

Your petitioner, of ———, ———, prays that he may be allowed to surrender the letters-patent for an improvement in coal-scuttles, granted to him May 16, 18 , whereof he is now sole owner, [or, " whereof C. D., on whose behalf and with whose assent this application is made, is now sole owner, by assignment,"] and that letters-patent may be reissued to him, [or, " the said C. D.,"] for the same invention, upon the annexed amended specification. Accompanying this petition is an abstract of title, duly certified, as required in such cases.
   A. B.

Assent of Assignee to Reissue.

The undersigned, assignee of the entire [or an undivided] interest in the above-mentioned letters-patent, hereby assents to the accompanying application.   C. D.

For a Reissue, (by Assignee.)

(To be used only when the inventor is dead, or the original patent was issued and assigned prior to July 8, 1870.)

To the Commissioner of Patents:

Your petitioners, of the city of ———, State of ———, pray that they may be allowed to surrender the letters-patent for an improvement in coal-scuttles, granted May 16, 18 , to E. F., now deceased, whereof they are now owners, by assignment, of the entire interest, and the letters-patent may be reissued to them for the same invention, upon the annexed amended specification. Accompanying this petition is an abstract of title, duly certified, as required in such cases.   A. B.
   C. D.

For Letters-Patent for a Design.

To the Commissioner of Patents:

Your petitioner, residing in ———, ———, prays that letters-patent may be granted to him for the term of three and one half years [or "seven years," or "fourteen years"] for the new and original design set forth in the annexed specification.   A. B.

For the Registration of a Trade-Mark.

To the Commissioner of Patents:
Your petitioners respectfully represent that the firm of A. B., C. D. & Co. is engaged in the manufacture of woven fabrics at ———, ———, and at ———, ———, and that the said firm is entitled to the exclusive use, upon the class of goods which they manufacture, of the trade-mark described in the annexed statement or specification, [and accompanying *fac-simile.*]
They therefore pray that they may be permitted to obtain protection for such lawful trade-mark under the law in such cases made and provided.

A. B., C. D. & Co.,
By A. B.

Petition with Power of Attorney.

To the Commissioner of Patents:
Your petitioner, a resident of the city of ———, State of ———, prays that letters-patent may be granted to him for the invention set forth in the annexed specification; and he hereby appoints C. D., of the city of ———, State of ———, his attorney, with full power of substitution and revocation, to prosecute this application, to make alterations and amendments therein, to receive the patent, and to transact all business in the Patent Office connected therewith.

A. B.

Power of Attorney.

If the power of attorney be given at any time other than that of making application for patent, it will be in substantially the following form:

To the Commissioner of Patents:
The undersigned having, on or about the 20th day of July, 18 , made application for letters-patent for an improvement in a· horse-power hereby appoints C. D., of the city of ———, State of ———, his attorney, with full power of substitution and revocation, to prosecute said application, to make alterations and amendments therein, to receive the patent, and to transact all business in the Patent Office connected therewith.
Signed at ———, and State of ———, this ——— day of ———, 18—.

A. B.

FORMS. 63

Revocation of Power of Attorney.
The undersigned having, on or about the 26th day of December, 18 , appointed C. D., of the city of ———, and State of ———, his attorney to prosecute an application for letters patent, made on or about the 1st day of June, 1868, for an improvement in the running-gear of wagons, hereby revokes the power of attorney then given.
Signed at ———, ———, this 21st day of July, 18 .
A. B.

SPECIFICATION.

To all whom it may concern:
Be it known that I, [here insert the name of the inventor,] of ———, in the county of ———, and State of ———, have invented a new and useful improvement in saw-toothing machines, which improvement is fully set forth in the following specification, reference being had to the accompanying drawings:

The object of my invention is to rapidly form, on the blade of a hand-saw, teeth gradually decreasing in size from the broad to the narrow end of the blade, by the combination, in a saw-toothing machine, of a tapering barrel, E, and a chain, or its equivalent, with rollers, $a\ a^1$, for feeding, or with a slide for carrying the blade A, as shown in the perspective view, Fig. 1, of the accompanying drawing.

The machine is illustrated more in detail in the plane view, Fig. 3, and in the vertical section, Fig. 2, in which it has not been deemed necessary to show the driving mechanism. The blade is held by and between the two upper rollers $a\ a^1$ (the latter being a feed roller), and two lower rollers $b\ b^1$, and is made to traverse in the direction of the arrow, at a gradually decreasing speed, by causing a barrel, D, to unwind a chain or its equivalent from a tapering barrel, E, on the shaft B. The several shafts have their bearings in a simple frame, H, the front portion h of the latter forming a table, which, in conjunction with the lower rollers, supports the blade, as the latter is caused to traverse with its edges in contact with the adjustable guides $y\ y$, on the frame. In this table is a fixed die or anvil, f, on which the blade bears, and in which is a triangular notch, corresponding in shape to a punch, e, on a rapidly revolving disk, G.

As the blade moves at a gradually decreasing speed in the direction of the arrow, the punch will strike triangular pieces from its edge, and the result will be the formation of the desired graduated teeth.

It will be evident that the driving-barrel, D, may be tapering, and the barrel, E, cylindrical, or that both barrels may be tapering, and arranged to feed gradually faster instead of gradually slower, with the same result, and that the blade may be clamped to a guided sliding-bed, controlled by a tapering barrel and cord or chain.

I claim as my invention—The combination in a saw-toothing machine, substanially as described, of a tapering barrel and chain, with a roller for feeding the blade. A. B.
Witnesses: C. D. and E. F.

Affidavit of Invention. By a Sole Inventor.
(To follow Specification.)

STATE OF ———, *County of* ———, *ss.*:—A. B., the above-named petitioner, being duly sworn (or affirmed), deposes and says that he verily believes himself to be the original and first inventor of the improvement in seed-drills described and claimed in the foregoing specification; that the same has not been patented to him or to others, with his knowledge and consent, in any country; that the same has not to his knowledge been in public use, or on sale in the United States, for more than two years prior to this application; that he does not know and does not believe that the same was ever before known or used; and that he is a citizen of ———, and a resident of ———. A. B.

Sworn to and subscribed before me this 13th day of March, 18 . C. D., *Justice of the Peace*.

[If the applicant be an alien, the sentence "and that he is a citizen of the United States" will be omitted, and in lieu thereof will be substituted "and that he is a citizen of the Republic of Mexico," or "and that he is a subject of the King of Italy," or "of the Queen of Great Britain," or as the case may be.

If the applicants claim to be *joint inventors*, the oath will read "that they verily believe themselves to be the original, first, and joint inventors," etc.

If the inventor be dead, the oath will be taken by the administrator or executor, and will declare his belief that the party named as inventor was the original and first inventor.]

Trade-Mark Declaration.

STATE OF———, *County of*———, *ss.*:
———, a resident of———, in the County of———, and State of———, being duly sworn, deposes and says that he is the applicant named in the foregoing statement; that he verily believes that the foregoing statement is true; that he has at this time a right to the use of the trade-mark herein described; that no other person, firm or corporation has the right to such use, either in the identical form or in any such near resemblance thereto as might be calculated to deceive; that it is used by him in commerce between the United

States and *———; that the description and fac-similes presented for record truly represent the trade-mark sought to be registered, and that he is a citizen of the United States.

Applicant signs here. ☞ ———

Sworn and subscribed before me, a ———, this ——— day of ———, 18—

Seal here—to be } (Signature of Justice, Notary or U. S. Consul.) ☞———
impressed in }
the paper. } (Official Character.) ———

* Herein must be inserted the name or names of foreign nation, nations or Indian tribes with which you are trading.

PETITION FOR CAVEAT.

The petition of A. B., of ———, in the county of ———, and State of ———, respectfully represents:

That he has made certain improvements in velocipedes, and that he is now engaged in making experiments for the purpose of perfecting the same, preparatory to applying for letters-patent therefor. He therefore prays that the subjoined description of his invention may be filed as a caveat in the confidential archives of the Patent Office.   A. B.

SPECIFICATION (*for Caveat*).

The following is a description of my newly-invented velocipede, which is as full, clear, and exact as I am able at this time to give, reference being had to the drawing hereto annexed.

This invention relates to that class of velocipedes in which there are two wheels connected by a beam forming a saddle for the rider, the feet being applied to cranks that revolve the front wheel.

The object of my invention is to render it unnecessary to turn the front wheel so much as heretofore, and at the same time to facilitate the turning of sharp curves. This I accomplish by fitting the front and the hind wheels on vertical pivots, and connecting them by means of a diagonal bar, as shown in the drawing, so that the turning of the front wheel also turns the back wheel with a position at an angle with the beams, thereby enabling it easily to turn a curve.

In the drawing, A is the front wheel, B the hind wheel, and C the standards extending from the axle of the front wheel to the vertical pivot a in the beam b, and D is the cross-bar upon the end of a, by which the steering is done. The hind wheel B is also fitted with jaws c and a vertical pivot, d.

Witnesses: C. D.   A. B.
              E. F.

[The form of oath will be substantially that provided for original applications, except that, as a caveat can only be filed by a citizen, or an alien who has resided for one year last past in the United States, and made oath of his intention to become a citizen, the oath should be modified accordingly.]

ASSIGNMENTS.

Of an undivided Fractional Interest in an Invention before the Issue of Letters-Patent.

In consideration of one dollar, to me paid by C. D., of ———, I do hereby sell and assign to said C. D. an undivided half of all my right, title, and interest in and to a certain invention in ploughs, as fully set forth and described in the specification which I have prepared [if the application has been already made, say "and filed"] preparatory to obtaining letters-patent of the United States therefor. And I do hereby authorize and request the Commissioner of Patents to issue the said letters-patent jointly to myself and the said C. D., our heirs and assigns.

Witness my hand and seal this   th day of February, 18
In presence of   A. B.

Of the Entire Interest in Letters-Patent.

In consideration of five hundred dollars, to me paid by C. D., of ———, I do hereby sell and assign to the said C. D. all my right, title, and interest in and to the letters-patent of the United States No. 41,806, for an improvement in locomotive head-lights, granted to me July 30, 18 , the same to be held and enjoyed by the said C. D. to the full end of the term for which said letters-patent are granted, as fully and entirely as the same would have been held and enjoyed by me if this assignment and sale had not been made.

Witness my hand and seal this   th day of January, 18 .
In presence of   A. B.

Of an Undivided Interest in the Letters-Patent and Extension thereof.

In consideration of one thousand dollars, to me paid by C. D., of ———, I do hereby sell and assign to the said C. D. one undivided fourth part of all my right, title, and interest in and to the letters-patent of the United States No. 10,485, for an improvement in cooking-stoves, granted to me May 16, 18 ; the same to be held and enjoyed by the said C. D. to the full end of the term for which said letters-patent are granted, and for the term of any extension thereof, as fully and entirely as the same would have been held and enjoyed by me if this assignment and sale had not been made.

Witness my hand and seal this   th day of June, 18 .
In presence of   A. B.

FORMS. 67

Exclusive Territorial Grant by an Assignee.

In consideration of one thousand dollars, to me paid by C. D., of ———, I do hereby grant and convey to the said C. D. the exclusive right to make, use, and vend within the State of ———, and in no other place or places, the improvement in corn-planters for which letters-patent of the United States, dated August 15, 18 , were granted to E. F., and by said E. F. assigned to me December 3, 18 , by an assignment duly recorded in liber X^8, p. 416, of the records of the Patent Office, the same to be held and enjoyed by the said C. D. as fully and entirely as the same would have been held and enjoyed by me if this grant had not been made.

Witness my hand and seal this   th day of March, 18 .
A. B.

License—Shop-Right.

In consideration of fifty dollars to me paid by the firm of S. J. & Co., of ———, I do hereby license and empower the said S. J. & Co. to manufacture, at a single foundry and machine-shop in said ———, and in no other place or places, the improvement in cotton-seed planters for which letters-patent of the United States No. 71,846 were granted to me November 13, 18 , and to sell the machines so manufactured throughout the United States to the full end of the term for which said letters-patent are granted.

Witness my hand and seal this   d day of April, 18 .
A. B.

License—not exclusive—with Royalty.

This agreement, made this 12th day of September, 187 , between A. B., of ———, party of the first part, and the Uniontown Agricultural Works of ———, party of the second part, witnesseth, that, whereas letters-patent of the United States for an improvement in horse-rakes were granted to the party of the first part, dated October 4, 18 ; and whereas the party of the second part is desirous of manufacturing horse-rakes containing said patented improvement: now, therefore, the parties have agreed as follows:

I. The party of the first part hereby licenses and empowers the party of the second part to manufacture, subject to the conditions hereinafter named, at their factory in ———, and in no other place or places, to the end of the term for which said letters-patent were granted, horse-rakes containing the patented improvements, and to sell the same within the United States.

II. The party of the second part agrees to make full and

true returns to the party of the first part, under oath, upon the first days of July and January in each year, of all horse-rakes containing the patented improvements manufactured by them.

III. The party of the second part agrees to pay to the party of the first part five dollars, as a license-fee upon every horse-rake manufactured by said party of the second part containing the patented improvements; provided, that if the said fee be paid upon the days provided herein for semi-annual returns, or within ten days thereafter, a discount of fifty per cent shall be made from said fee for prompt payment.

IV. Upon a failure of the party of the second part to make returns, or to make payment of license-fees, as herein provided, for thirty days after the days herein named, the party of the first part may terminate this license by serving a written notice upon the party of the second part; but the party of the second part shall not thereby be discharged from any liability to the party of the first part for any license-fees due at the time of the service of said notice.

In witness whereof the parties above named (the said Uniontown Agricultural Works, by its president) have hereunto set their hands the day and year first above written.

In presence of   A. B.
                                     U. A. W.

Transfer of a Trade-Mark.

We, A. B. and C. D., of ———, partners under the firm name of B. & D., in consideration of five hundred dollars to us paid by E. F., of the same place, do hereby sell, assign, and transfer to the said E. F. and his assigns the exclusive right to use in the manufacture and sale of stoves a certain trade-mark for stoves deposited by us in the United States Patent Office, and recorded therein July 15, 1870; the same to be held, enjoyed, and used by the said E. F., as fully and entirely as the same would have been held and enjoyed by us if this grant had not been made.

Witness our hands this 20th day of July, 187 .

In presence of   A. B.
                                     C. D.

Messrs. MUNN & CO , 361 Broadway, New York, make it a part of their business to prepare and put on record assignments and agreements relating to patents, trade-marks, copyrights, etc.

THE
PATENT LAWS
OF THE
UNITED STATES.

REVISED STATUTES, FORTY-THIRD CONGRESS, APPROVED JUNE 22, 1874.

ORGANIZATION OF THE PATENT OFFICE—SALARIES—POWERS OF THE COMMISSIONER—PRINTING OF PATENTS, ETC.

TITLE XI., Rev. Stat., sec. 440, p. 74:

There shall be in the Department of the Interior

In the Patent Office:

One chief clerk, at a salary of two thousand five hundred dollars a year.

One examiner in charge of interferences, at a salary of two thousand five hundred dollars a year.

One examiner in charge of trade-marks, at a salary of two thousand five hundred dollars a year.

Twenty-four principal examiners, at a salary of two thousand five hundred dollars a year each.

Twenty-four first assistant examiners, at a salary of one thousand eight hundred dollars a year each.

Twenty-four second assistant examiners (two of whom may be women), at a salary of one thousand six hundred dollars a year each.

Twenty-four third assistant examiners, at a salary of one thousand four hundred dollars a year each.

One librarian, at a salary of two thousand dollars a year.

One machinist, at a salary of one thousand six hundred dollars a year.

Three skilled draughtsmen, at a salary of one thousand two hundred dollars a year each.

Thirty-five copyists of drawings, at a salary of one thousand dollars a year each.

One messenger and purchasing clerk, at a salary of one thousand dollars a year.

One skilled laborer, at a salary of one thousand two hundred dollars a year.

Eight attendants in the model-room, at a salary of one thousand dollars a year each.

Eight attendants in the model-room, at a salary of nine hundred dollars a year each.

SEC. 475. There shall be in the Department of the Interior an office known as the Patent Office, where all records, books, models, drawings, specifications, and other papers and things pertaining to patents shall be safely kept and preserved.

SEC. 476. There shall be in the Patent Office a Commissioner of Patents, one Assistant Commissioner, and three examiners-in-chief, who shall be appointed by the President, by and with the advice and consent of the Senate. All other officers, clerks, and employés authorized by law for the Office shall be appointed by the Secretary of the Interior, upon the nomination of the Commissioner of Patents.

SEC. 477. The salaries of the officers mentioned in the preceding section shall be as follows:

The Commissioner of Patents, four thousand five hundred dollars a year.

The Assistant Commissioner of Patents, three thousand dollars a year.

Three examiners-in-chief, three thousand dollars a year each.

SEC. 478. The seal heretofore provided for the Patent Office shall be the seal of the Office, with which letters-patent and papers issued from the Office shall be authenticated.

SEC. 479. The Commissioner of Patents and the chief clerk, before entering upon their duties, shall severally give bond, with sureties, to the Treasurer of the United States, the former in the sum of ten thousand dollars, and the latter in the sum of five thousand dollars, conditioned for the faithful discharge of their respective duties, and that they shall render to the proper officers of the Treasury a true account of all money received by virtue of their offices.

SEC. 480. All officers and employés of the Patent Office shall be incapable, during the period for which they hold their appointments, to acquire or take, directly or indirectly, except by inheritance or bequest, any right or interest in any patent issued by the Office.

SEC. 481. The Commissioner of Patents, under the direction of the Secretary of the Interior, shall superintend or perform all duties respecting the granting and issuing of patents directed by law; and he shall have charge of all books, records, papers, models, machines, and other things belonging to the Patent Office.

SEC. 482. The examiners-in-chief shall be persons of competent legal knowledge and scientific ability, whose duty it shall be, on the written petition of the appellant, to revise and determine upon the validity of the adverse decisions of examiners upon applications for patents, and for reissues of patents, and in interference cases; and, when required by the Commissioner, they shall hear and report upon claims for extensions, and perform such other like duties as he may assign them.

SEC. 483. The Commissioner of Patents, subject to the approval of the Secretary of the Interior, may from time to time establish regulations, not inconsistent with law, for the conduct of proceedings in the Patent Office.

SEC. 484. The Commissioner of Patents shall cause to be classified and arranged in suitable cases, in the rooms and galleries provided for that purpose, models, specimens of composition, fabrics, manufactures, works of art, and designs, which have been or shall be deposited in the Patent Office; and the rooms and galleries shall be kept open during suitable hours for public inspection.

SEC. 485. The Commissioner of Patents may restore to the respective applicants such of the models belonging to rejected applications as he shall not think necessary to be preserved, or he may sell or otherwise dispose of them after the application has been finally rejected for one year, paying the proceeds into the Treasury, as other patent moneys are directed to be paid.

SEC. 486. There shall be purchased for the use of the Patent Office a library of such scientific works and periodicals, both foreign and American, as may aid the officers in the discharge of their duties, not exceeding the amount annually appropriated for that purpose.

SEC. 487. For gross misconduct the Commissioner of Patents may refuse to recognize any person as a patent agent, either generally or in any particular case; but the reasons for such refusal shall be duly recorded, and be subject to the approval of the Secretary of the Interior.

SEC. 488. The Commissioner of Patents may require all papers filed in the Patent Office, if not correctly, legibly, and clearly written, to be printed at the cost of the party filing them.

SEC. 489. The Commissioner of Patents may print, or

cause to be printed, copies of the claims of current issues, and copies of such laws, decisions, regulations, and circulars as may be necessary for the information of the public.

SEC. 490. The Commissioner of Patents is authorized to have printed, from time to time, for gratuitous distribution, not to exceed one hundred and fifty copies of the complete specifications and drawings of each patent hereafter issued, together with suitable indexes, one copy to be placed for free public inspection in each capitol of every State and Territory, one for the like purpose in the clerk's office of the district court of each judicial district of the United States, except when such offices are located in State or Territorial capitols, and one in the Library of Congress, which copies shall be certified under the hand of the Commissioner and seal of the Patent Office, and shall not be taken from the depositories for any other purpose than to be used as evidence.

SEC. 491. The Commissioner of Patents is authorized to have printed such additional numbers of copies of specifications and drawings, certified as provided in the preceding section, at a price not to exceed the contract price for such drawings, for sale, as may be warranted by the actual demand for the same; and he is also authorized to furnish a complete set of such specifications and drawings to any public library which will pay for binding the same into volumes to correspond with those in the Patent Office, and for the transportation of the same, and which shall also provide for proper custody for the same, with convenient access for the public thereto, under such regulations as the Commissioner shall deem reasonable.

SEC. 492. The lithographing and engraving required by the two preceding sections shall be awarded to the lowest and best bidders for the interest of the Government, due regard being paid to the execution of the work, after due advertising by the Congressional Printer under the direction of the Joint Committee on Printing; but the Joint Committee on Printing may empower the Congressional Printer to make immediate contracts for engraving, whenever, in their opinion, the exigencies of the public service will not justify waiting for advertisement and award; or if, in the judgment of the Joint Committee on Printing, the work can be performed under the direction of the Commissioner of Patents more advantageously than in the manner above prescribed, it shall be so done, under such limitations and conditions as the Joint Committee on Printing may from time to time prescribe.

SEC. 493. The price to be paid for uncertified printed

copies of specifications and drawings of patents shall be determined by the Commissioner of Patents, within the limits of ten cents as the minimum and fifty cents as the maximum price.

SEC. 494. The Commissioner of Patents shall lay before Congress, in the month of January, annually, a report, giving a detailed statement of all moneys received for patents, for copies of records or drawings, or from any other source whatever; a detailed statement of all expenditures for contingent and miscellaneous expenses; a list of all patents which were granted during the preceding year, designating under proper heads the subjects of such patents; an alphabetical list of all the patentees, with their places of residence; a list of all patents which have been extended during the year; and such other information of the condition of the Patent Office as may be useful to Congress or the public.

SEC. 495. The collections of the Exploring Expedition, now in the Patent Office, shall be under the care and management of the Commissioner of Patents.

SEC. 496. All disbursements for the Patent Office shall be made by the disbursing clerk of the Interior Department.

TITLE XIII., Rev. Stat., p. 168:

SEC. 571. [Refers to jurisdiction of certain district courts.]

SEC. 629. Clause 9. [The circuit courts have original jurisdiction] of all suits at law or in equity arising under the patent or copyright laws of the United States.

SEC. 892. Written or printed copies of any records, books, papers, or drawings belonging to the Patent Office, and of letters patent authenticated by the seal and certified by the Commissioner or Acting Commissioner thereof, shall be evidence in all cases wherein the originals could be evidence; and any person making application therefor, and paying the fee required by law, shall have certified copies thereof.

SEC. 893. Copies of the specifications and drawings of foreign letters-patent, certified as provided in the preceding section, shall be prima-facie evidence of the fact of the granting of such letters-patent, and of the date and contents thereof.

SEC. 894. The printed copies of specifications and drawings of patents, which the Commissioner of Patents is authorized to print for gratuitous distribution, and to deposit in the capitols of the States and Territories, and in the clerk's offices of the district courts, shall, when certified by him and authenticated by the seal of his office, be received in all courts as evidence of all matters therein contained.

TITLE XV., Rev. Stat., p. 261:

SEC. 1537. No patented article connected with marine engines shall hereafter be purchased or used in connection with any steam vessels of war until the same shall have been submitted to a competent board of naval engineers, and recommended by such board, in writing, for purchase and use.

TITLE XVII., Rev. Stat., p. 292:

SEC. 1672. The breech-loading system for muskets and carbines adopted by the Secretary of War, known as "the Springfield breech-loading system," is the only system to be used by the Ordnance Department, in the manufacture of muskets and carbines for the military service.

SEC. 1673. No royalty shall be paid by the United States to any one of its officers or employés for the use of any patent for the system, or any part thereof, mentioned in the preceding section, nor for any such patent in which said officers or employés may be directly or indirectly interested.

Concerning Applications for, and Issue of, Patents.

TITLE LX., Rev. Stat., chap. 1, p. 953:

SEC. 4883. All patents shall be issued in the name of the United States of America, under the seal of the Patent Office, and shall be signed by the Secretary of the Interior and countersigned by the Commissioner of Patents, and they shall be recorded, together with the specifications, in the Patent Office, in books to be kept for that purpose.

SEC. 4884. Every patent shall contain a short title or description of the invention or discovery, correctly indicating its nature and design, and a grant to the patentee, his heirs or assigns, for the term of seventeen years, of the exclusive right to make, use, and vend the invention or discovery throughout the United States, and the Territories thereof, referring to the specification for the particulars thereof. A copy of the specification and drawings shall be annexed to the patent and be a part thereof.

SEC. 4885. Every patent shall bear date as of a day not later than six months from the time at which it was passed and allowed and notice thereof was sent to the applicant or his agent; and if the final fee is not paid within that period, the patent shall be withheld.

SEC. 4886. Any person who has invented or discovered any new and useful art, machine, manufacture or composition of matter, or any new and useful improvement thereof, not known nor used by others in this country, and not patented or described in any printed publication in this or any foreign country, before his invention or discovery thereof, and not in public use or on sale for more than two years prior to his application, unless the same is proved to have been abandoned, may upon payment of the fees required by law, and other due proceedings had, obtain a patent therefor.

SEC. 4887. No person shall be debarred from receiving a patent for his invention or discovery, nor shall any patent be declared invalid, by reason of its having been first patented or caused to be patented in a foreign country, unless the same has been introduced into public use in the United States for more than two years prior to the application. But every patent granted for an invention which has been previously patented in a foreign country shall be so limited as to expire at the same time with the foreign patent, or, if there be more than one, at the same time with the one having the shortest term, and in no case shall it be in force more than seventeen years.

SEC. 4888. Before any inventor or discoverer shall receive a patent for his invention or discovery, he shall make application therefor, in writing, to the Commissioner of Patents, and shall file in the Patent Office a written description of the same, and of the manner and process of making, constructing, compounding, and using it, in such full, clear, concise, and exact terms as to enable any person skilled in the art or science to which it appertains, or with which it is most nearly connected, to make, construct, compound, and use the same; and in case of a machine, he shall explain the principle thereof, and the best mode in which he has contemplated applying that principle, so as to distinguish it from other inventions; and he shall particularly point out and distinctly claim the part, improvement, or combination which he claims as his invention or discovery. The specification and claim shall be signed by the inventor and attested by two witnesses.

SEC. 4889. When the nature of the case admits of drawings, the applicant shall furnish one copy signed by the inventor or his attorney in fact, and attested by two witnesses, which shall be filed in the Patent Office; and a copy of the drawing, to be furnished by the Patent Office, shall be attached to the patent as a part of the specification.

SEC. 4890. When the invention or discovery is of a composition of matter, the applicant, if required by the Commissioner, shall furnish specimens of ingredients and of the composition, sufficient in quantity for the purpose of experiment.

SEC. 4891. In all cases which admit of representation by

model, the applicant, if required by the Commissioner, shall furnish a model of convenient size to exhibit advantageously the several parts of his invention or discovery.

SEC. 4892. The applicant shall make oath that he does verily believe himself to be the original and first inventor or discoverer of the art, machine, manufacture, composition, or improvement for which he solicits a patent ; that he does not know and does not believe that the same was ever before known or used ; and shall state of what country he is a citizen. Such oath may be made before any person within the United States authorized by law to administer oaths, or when the applicant resides in a foreign country, before any minister, chargé d'affaires, consul, or commercial agent, holding commission under the Government of the United States, or before any notary public of the foreign country in which the applicant may be.

SEC. 4893. On the filing of any such application and the payment of the fees required by law. the Commissioner of Patents shall cause an examination to be made of the alleged new invention or discovery; and if on such examination it shall appear that the claimant is justly entitled to a patent under the law, and that the same is sufficiently useful and important, the Commissioner shall issue a patent therefor.

SEC. 4894. All applications for patents shall be completed and prepared for examination within two years after the filing of the application, and in default thereof, or upon failure of the applicant to prosecute the same within two years after any action therein, of which notice shall have been given to the applicant, they shall be regarded as abandoned by the parties thereto, unless it be shown to the satisfaction of the Commissioner of Patents that such delay was unavoidable.

Patents may be issued to Assignees.

SEC. 4895. Patents may be granted and issued or reissued to the assignee of the inventor or discoverer ; but the assignment must first be entered of record in the Patent Office. And in all cases of an application by an assignee for the issue of a patent, the application shall be made and the specification sworn to by the inventor or discoverer; and in all cases of an application for a reissue of any patent, the application must be made and the corrected specification signed by the inventor or discoverer, if he is living, unless the patent was issued and the assignment made before the eighth day of July, eighteen hundred and seventy.

Patents to the Heirs of Deceased Inventors.

SEC. 4896. When any person, having made any new invention or discovery for which a patent might have been granted, dies before a patent is granted, the right of applying for and obtaining the patent shall devolve on his executor or administrator, in trust for the heirs at law of the deceased, in case he shall have died intestate; or if he shall have left a will, disposing of the same, then in trust for his devisees, in as full manner and on the same terms and conditions as the same might have been claimed or enjoyed by him in his lifetime; and when the application is made by such legal representatives, the oath or affirmation required to be made shall be so varied in form that it can be made by them.

Renewal of Lapsed Cases.

SEC. 4897. Any person who has an interest in an invention or discovery, whether as inventor, discoverer, or assignee, for which a patent was ordered to issue upon the payment of the final fee, but who fails to make payment thereof within six months from the time at which it was passed and allowed and notice thereof was sent to the applicant or his agent, shall have a right to make an application for a patent for such invention or discovery the same as in the case of an original application. But such second application must be made within two years after the allowance of the original application. But no person shall be held responsible in damages for the manufacture or use of any article or thing for which a patent was ordered to issue under such renewed application prior to the issue of the patent. And upon the hearing of renewed applications preferred under this section, abandonment shall be considered as a question of fact.

Assignment of Patents.

SEC. 4898. Every patent or any interest therein shall be assignable in law by an instrument in writing; and the patentee or his assigns or legal representatives may, in like manner, grant and convey an exclusive right under his patent to the whole or any specified part of the United States. An assignment, grant, or conveyance shall be void as against any subsequent purchaser or mortgagee for a valuable consideration, without notice, unless it is recorded in the Patent Office within three months from the date thereof.

Free Rights of Use.

SEC. 4899. Every person who purchases of the inventor, or discoverer, or with his knowledge and consent constructs any newly invented or discovered machine, or other patentable article, prior to the application by the inventor or discoverer for a patent, or who sells or uses one so constructed, shall have the right to use, and vend to others to be used, the specific thing so made or purchased, without liability therefor.

Patented Articles to be Stamped.

SEC. 4900. It shall be the duty of all patentees, and their assigns and legal representatives, and of all persons making or vending any patented article for or under them, to give sufficient notice to the public that the same is patented; either by affixing thereon the word "patented," together with the day and year the patent was granted; or when, from the character of the article, this can not be done, by fixing to it, or to the package wherein one or more of them is inclosed, a label containing the like notice; and in any suit for infringement, by the party failing so to mark, no damages shall be recovered by the plaintiff, except on proof that the defendant was duly notified of the infringement, and continued, after such notice, to make, use, or vend the article so patented.

Penalties for False Stamping.

SEC. 4901. Every person who, in any manner, marks upon any thing made, used, or sold by him for which he has not obtained a patent, the name or any imitation of the name of any person who has obtained a patent therefor, without the consent of such patentee, or his assigns or legal representatives; or

Who, in any manner, marks upon or affixes to any such patented article the word "patent" or "patentee," or the words "letters-patent," or any word of like import, with intent to imitate or counterfeit the mark or device of the patentee, without having the license or consent of such patentee or his assigns or legal representatives; or

Who, in any manner, marks upon or affixes to any unpatented article the word "patent" or any word importing that the same is patented, for the purpose of deceiving the public, shall be liable, for every such offence, to a penalty of not less than one hundred dollars, with costs; one half of said penalty to the person who shall sue for the same, and the other to the use of the United States, to be recovered by suit in any district court of the United States within whose jurisdiction such offence may have been committed.

Caveats.

Sec. 4902. Any citizen of the United States who makes any new invention or discovery, and desires further time to mature the same, may, on payment of the fees required by law, file in the Patent Office a caveat setting forth the design thereof, and of its distinguishing characteristics, and praying protection of his right until he shall have matured his invention. Such caveat shall be filed in the confidential archives of the office and preserved in secrecy, and shall be operative for the term of one year from the filing thereof; and if application is made within the year by any other person for a patent with which such caveat would in any manner interfere, the Commissioner shall deposit the description, specification, drawings, and model of such application in like manner in the confidential archives of the office, and give notice thereof, by mail, to the person by whom the caveat was filed. If such person desires to avail himself of his caveat, he shall file his description, specifications, drawings, and model within three months from the time of placing the notice in the post-office in Washington, with the usual time required for transmitting it to the caveator added thereto; which time shall be indorsed on the notice. An alien shall have the privilege herein granted, if he has resided in the United States one year next preceding the filing of his caveat, and has made oath of his intention to become a citizen.

Rejected Cases.

Sec. 4903. Whenever, on examination, any claim for a patent is rejected, the Commissioner shall notify the applicant thereof, giving him briefly the reasons for such rejection, together with such information and references as may be useful in judging of the propriety of renewing his application or of altering his specification; and if, after receiving such notice, the applicant persists in his claim for a patent, with or without altering his specifications, the Commissioner shall order a re-examination of the case.

Interferences.

Sec. 4904. Whenever an application is made for a patent which, in the opinion of the Commissioner, would interfere with any pending application, or with any unexpired patent, he shall give notice thereof to the applicants, or applicant and patentee, as the case may be, and shall direct the primary examiner to proceed to determine the question of priority of invention. And the Commissioner may issue a patent to the party who is adjudged the prior inventor, unless the adverse party appeals from the decision of the primary examiner, or of the board of examiners-in-chief, as the case may be, within such time, not less than twenty days, as the Commissioner shall prescribe.

Relating to Witnesses.

SEC. 4905. The Commissioner of Patents may establish rules for taking affidavits and depositions required in cases pending in the Patent Office, and such affidavits and depositions may be taken before any officer authorized by law to take depositions to be used in the courts of the United States, or of the State where the officer resides.

SEC. 4906. The clerk of any court of the United States, for any district or Territory wherein testimony is to be taken for use in any contested case pending in the Patent Office, shall, upon the application of any party thereto, or of his agent or attorney, issue a subpœna for any witness residing or being within such district or Territory, commanding him to appear and testify before any officer in such district or Territory authorized to take depositions and affidavits, at any time and place in the subpœna stated. But no witness shall be required to attend at any place more than forty miles from the place where the subpœna is served upon him.

SEC. 4907. Every witness duly subpœnaed and in attendance shall be allowed the same fees as are allowed to witnesses attending the courts of the United States.

SEC. 4908. Whenever any witness, after being duly served with such subpœna, neglects or refuses to appear, or after appearing refuses to testify, the judge of the court whose clerk issued the subpœna may, on proof of such neglect or refusal, enforce obedience to the process, or punish the disobedience as in other like cases. But no witness shall be guilty of contempt for disobeying such subpœna, unless his fees and travelling expenses in going to, returning from, and one day's attendance at the place of examination, are paid or tendered him at the time of the service of the subpœna; nor for refusing to disclose any secret invention or discovery made or owned by himself.

Appeals.

SEC. 4909. Every applicant for a patent or for the reissue for a patent, any of the claims of which have been twice rejected, and every party to an interference, may appeal from the decision of the primary examiner, or of the examiner in

PATENT LAWS OF 1874.

charge of interferences in such case, to the board of examiners-in-chief; having once paid the fee for such appeal.

SEC. 4910. If such party is dissatisfied with the decision of the examiners-in-chief, he may, on payment of the fee prescribed, appeal to the Commissioner in person.

SEC. 4911. If such party, except a party to an interference, is dissatisfied with the decision of the Commissioner, he may appeal to the Supreme Court of the District of Columbia, sitting in banc.

SEC. 4912. When an appeal is taken to the Supreme Court of the District of Columbia, the appellant shall give notice thereof to the Commissioner, and file in the Patent Office, within such time as the Commissioner shall appoint, his reasons of appeal, specifically set forth in writing.

SEC. 4913. The court shall, before hearing such appeal, give notice to the Commissioner of the time and place of the hearing, and on receiving such notice the Commissioner shall give notice of such time and place in such manner as the court may prescribe, to all parties who appear to be interested therein. The party appealing shall lay before the court certified copies of all the original papers and evidence in the case, and the Commissioner shall furnish the court with the grounds of his decision, fully set forth in writing, touching all the points involved by the reasons of appeal. And at the request of any party interested, or of the court, the Commissioner and the examiners may be examined under oath, in explanation of the principles of the thing for which a patent is demanded.

SEC. 4914. The court, on petition, shall hear and determine such appeal, and revise the decision appealed from in a summary way, on the evidence produced before the Commissioner, at such early and convenient time as the court may appoint; and the revision shall be confined to the points set forth in the reasons of appeal. After hearing the case the court shall return to the Commissioner a certificate of its proceedings and decision, which shall be entered of record in the Patent Office, and shall govern the further proceedings in the case. But no opinion or decision of the court in any such case shall preclude any person interested from the right to contest the validity of such patent in any court wherein the same may be called in question.

SEC. 4915. Whenever a patent on application is refused, either by the Commissioner of Patents or by the Supreme Court of the District of Columbia upon appeal from the Commissioner, the applicant may have remedy by bill in equity; and the court having cognizance thereof, on notice to adverse parties and other due proceedings had, may adjudge that such applicant is entitled, according to law, to receive a patent for his invention, as specified in his claim, or for any part thereof, as the facts in the case may appear. And such adjudication, if it be in favor of the right of the applicant, shall authorize the Commissioner to issue such patent on the applicant filing in the Patent Office a copy of the adjudication, and otherwise complying with the requirements of law. In all cases, where there is no opposing party, a copy of the bill shall be served on the Commissioner; and all the expenses of the proceeding shall be paid by the applicant, whether the final decision is in his favor or not.

REISSUES.

SEC. 4916. Whenever any patent is inoperative or invalid, by reason of a defective or insufficient specification, or by reason of the patentee claiming as his own invention or discovery more than he had a right to claim as new, if the error has arisen by inadvertence, accident, or mistake, and without any fraudulent or deceptive intention, the Commissioner shall, on the surrender of such patent and the payment of the duty required by law, cause a new patent for the same invention, and in accordance with the corrected specification, to be issued to the patentee, or, in the case of his death or of an assignment of the whole or any undivided part of the original patent, then to his executors, administrators, or assigns, for the unexpired part of the term of the original patent. Such surrender shall take effect upon the issue of the amended patent. The Commissioner may, in his discretion, cause several patents to be issued for distinct and separate parts of the thing patented, upon demand of the applicant, and upon payment of the required fee for a reissue for each of such reissued letters-patent. The specifications and claim in every such case shall be subject to revision and restriction in the same manner as original applications are. Every patent so reissued, together with the corrected specification, shall have the same effect and operation in law, on the trial of all actions for causes thereafter arising, as if the same had been originally filed in such corrected form; but no new matter shall be introduced into the specification, nor in case of a machine patent shall the model or drawings be amended, except each by the other; but when there is neither model nor drawing, amendments may be made upon proof satisfactory to the Commissioner that such new matter or amendment was a part of the original invention, and was omitted from the specification by inadvertence, accident, or mistake, as aforesaid.

DISCLAIMERS.

SEC. 4917. Whenever, through inadvertence, accident, or mistake, and without any fraudulent or deceptive intention, a patentee has claimed more than that of which he was the original or first inventor or discoverer, his patent shall be valid for all that part which is truly and justly his own, provided the same is a material or substantial part of the thing patented; and any such patentee, his heirs or assigns, whether of the whole or any sectional interest therein, may, on payment of the fee required by law, make disclaimer of such parts of the thing patented as he shall not choose to claim or to hold by virtue of the patent or assignment, stating therein the extent of his interest in such patent. Such disclaimer shall be in writing, attested by one or more witnesses, and recorded in the Patent Office; and it shall thereafter be considered as part of the original specification to the extent of the interest possessed by the claimant and by those claiming under him after the record thereof. But no such disclaimer shall affect any action pending at the time of its being filed, except so far as may relate to the question of unreasonable neglect or delay in filing it.

INTERFERING PATENTS.

SEC. 4918. Whenever there are interfering patents, any person interested in any one of them, or in the working of the invention claimed under either of them, may have relief against the interfering patentee, and all parties interested under him, by suit in equity against the owners of the interfering patent; and the court, on notice to adverse parties, and other due proceedings had according to the course of equity, may adjudge and declare either of the patents void in whole or in part, or inoperative, or invalid in any particular part of the United States, according to the interest of the parties in the patent or the invention patented. But no such judgment or adjudication shall affect the right of any person except the parties to the suit and those deriving title under them subsequent to the rendition of such judgment.

INFRINGEMENTS.

SEC. 4919. Damages for the infringement of any patent may be recovered by action on the case, in the name of the party interested, either as patentee, assignee, or grantee. And whenever in any such action a verdict is rendered for the plaintiff, the court may enter judgment thereon for any sum above the amount found by the verdict as the actual damages sustained, according to the circumstances of the case, not exceeding three times the amount of such verdict, together with the costs.

SEC. 4920. In any action for infringement the defendant may plead the general issue, and having given notice in writing to the plaintiff or his attorney, thirty days before, may prove, on trial, any one or more of the following special matters:

First. That for the purpose of deceiving the public the description and specification filed by the patentee in the Patent Office was made to contain less than the whole truth relative to his invention or discovery, or more than is necessary to produce the desired effect; or,

Second. That he had surreptitiously or unjustly obtained the patent for that which was in fact invented by another, who was using reasonable diligence in adapting and perfecting the same; or,

Third. That it had been patented or described in some printed publication prior to his supposed invention or discovery thereof; or,

Fourth. That he was not the original and first inventor or discoverer of any material and substantial part of the thing patented; or,

Fifth. That it had been in public use or on sale in this country for more than two years before his application for a patent, or had been abandoned to the public.

And in notices as to proof of previous invention, knowledge, or use of the thing patented, the defendant shall state the names of patentees and the dates of their patents, and when granted, and the names and residences of the persons alleged to have invented, or to have had the prior knowledge of the thing patented, and where and by whom it had been used; and if any one or more of the special matters alleged shall be found for the defendant, judgment shall be rendered for him with costs. And the like defences may be pleaded in any suit in equity for relief against an alleged infringement; and proofs of the same may be given upon like notice in the answer of the defendant, and with the like effect.

SEC. 4921. The several courts vested with jurisdiction of cases arising under the patent laws shall have power to grant injunctions according to the course 'and principles of courts of equity, to prevent the violation of any right secured by patent, on such terms as the court may deem reasonable; and upon a decree being rendered in any such case for an infringement, the complainant shall be entitled to recover, in addition to the profits to be accounted for by the defendant, the damages the complainant has sustained thereby; and the court shall assess the same or cause the same to be assessed under its direction. And the court shall have the same power to increase such damages, in its discretion, as is given to increase the damages found by verdicts in actions in the nature of actions of trespass upon the case.

SEC. 4922. Whenever, through inadvertence, accident, or mistake, and without any wilful default or intent to defraud or mislead the public, a patentee has, in his specification, claimed to be the original and first inventor or discoverer of any material or substantial part of the thing patented, of which he was not the original and first inventor or discoverer, every such patentee, his executors, administrators, and assigns, whether of the whole or any sectional interest in the patent, may maintain a suit at law or in equity, for the infringement of any part thereof, which was *bona fide* his own, if it is a material and substantial part of the thing patented, and definitely distinguishable from the parts claimed without right, notwithstanding the specifications may embrace more than that of which the patentee was the first inventor or discoverer. But in every such case in which a judgment or decree shall be rendered for the plaintiff no costs shall be recovered unless the proper disclaimer has been entered at the Patent Office before the commencement of the suit. But no patentee shall be entitled to the benefits of this section if he has unreasonably neglected or delayed to enter a disclaimer.

SEC. 4923. Whenever it appears that a patentee, at the time of making his application for the patent, believed himself to be the original and first inventor or discoverer of the thing patented, the same shall not be held to be void on account of the invention or discovery, or any part thereof, having been known or used in a foreign country, before his invention or discovery thereof, if it had not been patented or described in a printed publication.

SEC. 4928. The benefit of the extension of a patent shall extend to the assignees and grantees of the right to use the thing patented, to the extent of their interest therein.

PATENTS FOR DESIGNS.*

SEC. 4929. Any person who, by his own industry, genius, efforts, and expense, has invented and produced any new and original design for a manufacture, bust, statue, alto-relievo, or bas-relief; any new and original design for the printing of woollen, silk, cotton, or other fabrics; any new and original impression, ornament, patent, [pattern,] print, or picture to be

---

*See page 144 for new law.

printed, painted, cast, or otherwise placed on or worked into any article of manufacture; or any new, useful, and original shape or configuration of any article of manufacture, the same not having been known or used by others before his invention or production thereof, or patented or described in any printed publication, may, upon payment of the fee prescribed, and other due proceedings had the same as in cases of inventions or discoveries, obtain a patent therefor.

SEC. 4930. The Commissioner may dispense with models of designs when the design can be sufficiently represented by drawings or photographs.

SEC. 4931. Patents for designs may be granted for the term of three years and six months, or for seven years, or for fourteen years, as the applicant may, in his application, elect.

SEC. 4932. Patentees of designs issued prior to the second day of March, eighteen hundred and sixty-one, shall be entitled to extension of their respective patents for the term of seven years, in the same manner and under the same restrictions as are provided for the extension of patents for inventions or discoveries, issued prior to the second day of March, eighteen hundred and sixty-one.

SEC. 4933. All the regulations and provisions which apply to obtaining or protecting patents for inventions or discoveries not inconsistent with the provisions of this title, shall apply to patents for designs. [See new law, page 144.]

OFFICIAL FEES.

SEC. 4934. The following shall be the rates for patent fees:

On filing each original application for a patent, except in design cases, fifteen dollars.

On issuing each original patent, except in design cases, twenty dollars.

In design cases: For three years and six months, ten dollars; for seven years, fifteen dollars; for fourteen years, thirty dollars.

On filing each caveat, ten dollars.

On every application for the reissue of a patent thirty dollars.

On filing each disclaimer, ten dollars.

On every application for the extension of a patent, fifty dollars.

On the granting of every extension of a patent, fifty dollars.

On an appeal for the first time from the primary examiners to the examiners-in-chief, ten dollars.

On every appeal from the examiners-in-chief to the Commissioner, twenty dollars.

For certified copies of patents and other papers, including certified printed copies, ten cents per hundred words.

For recording every assignment, agreement, power of attorney, or other paper, of three hundred words or under, one dollar; of over three hundred and under one thousand words, two dollars; of over one thousand words, three dollars.

For copies of drawings, the reasonable cost of making them.

SEC. 4935. Patent fees may be paid to the Commissioner of Patents, or to the Treasurer or any of the assistant treasurers of the United States, or to any of the designated depositaries, national banks, or receivers of public money, designated by the Secretary of the Treasury for that purpose; and such officer shall give the depositor a receipt or certificate of deposit therefor. All money received at the Patent Office, for any purpose, or from any source whatever, shall be paid into the Treasury as received, without any deduction whatever.

SEC. 4936. The Treasurer of the United States is authorized to pay back any sum or sums of money to any person who has through mistake paid the same into the Treasury, or to any receiver or depositary, to the credit of the Treasury, as for fees accruing at the Patent Office, upon a certificate thereof being made to the Treasurer by the Commissioner of Patents.

THE TRADE-MARK LAW OF THE UNITED STATES.

The original trade-mark laws of the United States, passed July 8, 1870, and embodied, with additions, in the U. S. Statutes of 1874, were, by a decision of the Supreme Court of the United States, delivered November 18, 1879, declared to be unconstitutional and void. March 3, 1881, Congress passed a new trade-mark law, the text of which is as follows:

AN ACT TO AUTHORIZE THE REGISTRATION OF TRADE-MARKS AND PROTECT THE SAME.

Be it enacted by the Senate and House of Representatives of the United States in Congress assembled, That owners of trade-marks used in commerce with foreign nations or with the Indian tribes, provided such owners shall be domiciled in the United States or located in any foreign country or tribes which, by treaty, convention, or law, affords similar privileges to citizens of the United States, may obtain registration of such trade-marks by complying with the following requirements:

First. By causing to be recorded in the Patent Office a statement specifying name, domicile, location, and citizenship of the party applying; the class of merchandise and the particular description of goods comprised in such class to which the particular trade-mark has been appropriated; a description of the trade-mark itself, with facsimiles thereof, and a statement of the mode in which the same is applied and affixed to goods and the length of time during which the trade-mark has been used.

Second. By paying into the Treasury of the United States the sum of twenty-five dollars, and complying with such regulations as may be prescribed by the Commissioner of Patents.

SEC. 2. That the application prescribed in the foregoing section must, in order to create any right whatever in favor of the party filing it, be accompanied by a written declaration verified by the person, or by a member of a firm, or by an officer of a corporation applying, to the effect that such party has at the time a right to the use of the trade-mark sought to be registered, and that no other person, firm, or corporation has the right to such use, either in the identical form or in any such near resemblance thereto as might be calculated to deceive; that such trade-mark is used in commerce with foreign nations or Indian tribes, as above indicated; and that the description and facsimiles presented for registry truly represent the trade-mark sought to be registered.

SEC. 3. That the time of the receipt of any such application shall be noted and recorded. But no alleged trade-mark shall be registered unless the same appear to be lawfully used as such by the applicant in foreign commerce or commerce with Indian tribes, as above mentioned, or is within the provision of a treaty, convention, or declaration with a foreign power; nor which is merely the name of the applicant; nor which is identical with a registered or known trade-mark owned by another and appropriate to the same class of merchandise, or which so nearly resembles some other person's lawful trade-mark as to be likely to cause confusion or mistake in the mind of the public, or to deceive purchasers. In an application for registration the Commissioner of Patents shall decide the presumptive lawfulness of claim to the alleged trade-mark; and in any dispute between an applicant and a previous registrant, or between applicants, he shall follow, so far as the same may be applicable, the practice of courts of equity of the United States in analogous cases.

SEC. 4. That certificates of registry of trade-marks shall

be issued in the name of the United States of America, under the seal of the Department of the Interior, and shall be signed by the Commissioner of Patents, and a record thereof, together with printed copies of the specifications, shall be kept in books for that purpose. Copies of trade-marks and of statements and declarations filed therewith and certificates of registry so signed and sealed, shall be evidence in any suit in which such trade-marks shall be brought in controversy.

SEC. 5. That a certificate of registry shall remain in force for thirty years from its date, except in cases where the trade-mark is claimed for and applied to articles not manufactured in this country, and in which it receives protection under the laws of a foreign country for a shorter period, in which case it shall cease to have any force in this country by virtue of this act at the time that such trade-mark ceases to be exclusive property elsewhere. At any time during the six months prior to the expiration of the term of thirty years such registration may be renewed on the same terms and for a like period.

SEC. 6. That applicants for registration under this act shall be credited for any fee or part of a fee heretofore paid into the Treasury of the United States with intent to procure protection for the same trade-mark.

SEC. 7. That registration of a trade-mark shall be *prima facie* evidence of ownership. Any person who shall reproduce, counterfeit, copy, or colorably imitate any trade-mark registered under this act and affix the same to merchandise of substantially the same descriptive properties as those described in the registration, shall be liable to an action on the case for damages for the wrongful use of said trade-mark at the suit of the owner thereof; and the party aggrieved shall also have his remedy according to the course of equity to enjoin the wrongful use of such trade-mark used in foreign commerce or commerce with Indian tribes as aforesaid, and to recover compensation therefor in any court having jurisdiction over the person guilty of such wrongful act; and courts of the United States shall have original and appellate jurisdiction in such cases without regard to the amount in controversy.

SEC. 8. That no action or suit shall be maintained under the provisions of this act in any case when the trade-mark is used in any unlawful business or upon any article injurious in itself, or which mark has been used with the design of deceiving the public in the purchase of merchandise, or under any certificate of registry fraudulently obtained.

SEC. 9. That any person who shall procure the registry of

a trade-mark, or of himself as the owner of a trade-mark, or an entry respecting a trade-mark, in the office of the Commissioner of Patents, by a false or fraudulent representation or declaration, orally or in writing, or by any fraudulent means, shall be liable to pay any damages sustained in consequence thereof to the injured party, to be recovered in an action on the case.

SEC. 10. That nothing in this act shall prevent, lessen, impeach, or avoid any remedy at law or in equity which any party aggrieved by any wrongful use of any trade-mark might have had if the provisions of this act had not been passed.

SEC. 11. That nothing in this act shall be construed as unfavorably affecting a claim to a trade-mark after the term of registration shall have expired; nor to give cognizance to any court of the United States in an action or suit between citizens of the same State, unless the trade-mark in controversy is used on goods intended to be transported to a foreign country, or in lawful commercial intercourse with an Indian tribe.

SEC. 12. That the Commissioner of Patents is authorized to make rules and regulations and prescribe forms for the transfer of the right to use trade-marks and for recording such transfers in his office.

SEC. 13. That citizens and residents of this country wishing the protection of trade-marks in any foreign country the laws of which require registration here as a condition precedent to getting such protection there, may register their trade-marks for that purpose as is above allowed to foreigners, and have certificate thereof from the Patent Office.

Approved March 3, 1881.

AN ACT

TO PUNISH THE COUNTERFEITING OF TRADE-MARK GOODS AND THE SALE OR DEALING IN OF COUNTERFEIT TRADE-MARK GOODS

Approved August 14th, 1876.

Be it enacted by the Senate and House of Representatives of the United States of America in Congress assembled, That every person who shall with intent to defraud, deal in or sell, or keep or offer for sale, or cause or procure the sale of, any goods of substantially the same descriptive properties as those referred to in the registration of any trade-mark, pursuant to the statutes of the United States, to which, or to the package in which the same are put up, is fraudulently affixed said trade-mark, or any colorable imitation thereof, calculated to deceive the public, knowing the same to be counterfeit or not the genuine goods referred to in said registration, shall, on conviction thereof, be punished by fine not exceeding one thousand dollars, or imprisonment not more than two years, or both such fine and imprisonment.

SEC. 2. That every person who fraudulently affixes, or causes or procures to be fraudulently affixed, any trade-mark registered pursuant to the statutes of the United States, or any colorable imitation thereof, calculated to deceive the public, to any goods, of substantially the same descriptive properties as those referred to in said registration, or to the package in which they are put up, knowing the same to be counterfeit, or not the genuine goods referred to in said registration, shall, on conviction thereof, be punished as prescribed in the first section of this act.

SEC. 3. That every person who fraudulently fills, or causes or procures to be fraudulently filled, any package to which is affixed any trade-mark, registered pursuant to the statutes of the United States, or any colorable imitation thereof, calculated to deceive the public, with any goods of substantially the same descriptive properties as those referred to in said registration, knowing the same to be counterfeit, or not the genuine goods referred to in said registration, shall, on conviction thereof, be punished as prescribed in the first section of this act.

SEC. 4. That any person or persons who shall, with intent to defraud any person or persons, knowingly and wilfully cast, engrave, or manufacture, or have in his, her, or their possession, or buy, sell, offer for sale, or deal in, any die or dies, plate or plates, brand or brands, engraving or engravings, on wood, stone, metal, or other substance, moulds, or any false representation, likeness, copy, or colorable imitation of any die, plate, brand, engraving, or mould of any private label, brand, stamp, wrapper, engraving on paper or other substance, or trade-mark, registered pursuant to the statutes of the United States, shall, upon conviction thereof, be punished as prescribed in the first section of this act.

SEC. 5. That any person or persons who shall, with intent to defraud any person or persons, knowingly and wilfully make, forge, or counterfeit, or have in his, her, or their possession, or buy, sell, offer for sale, or deal in, any representation, likeness, similitude, copy, or colorable imitation of any private label, brand, stamp, wrapper, engraving, mould, or trade-mark, registered pursuant to the statutes of the United States, shall, upon conviction thereof, be punished as prescribed in the first section of this act.

SEC. 6. That any person who shall, with intent to injure or defraud the owner of any trade-mark, or any other person lawfully entitled to use or protect the same, buy, sell, offer for sale, deal in or have in his possession any used or empty box, envelope, wrapper, case, bottle, or other package, to which is affixed, so that the same may be obliterated without substantial injury to such box or other thing aforesaid, any trade-mark, registered pursuant to the statutes of the United States, not so defaced, erased, obliterated, and destroyed as to prevent its fraudulent use, shall, on conviction thereof, be punished as prescribed in the first section of this act.

SEC. 7. That if the owner of any trade-mark, registered pursuant to the statutes of the United States, or his agent, make oath, in writing, that he has reason to believe, and does believe, that any counterfeit dies, plates, brands, engravings on wood, stone, metal, or other substance, or moulds, of his said registered trade-mark, are in the possession of any person, with intent to use the same for the purpose of deception and fraud, or makes such oaths that any counterfeits or colorable imitations of his said trade-mark, label, brand, stamp, wrapper, engraving on paper or other substance, or empty box, envelope, wrapper, case, bottle or other package, to which is affixed said registered trade-mark not so defaced, erased, obliterated, and destroyed as to prevent its fraudulent use, are in the possession of any person, with intent to use the same for the purpose of deception and fraud, then the several judges of the circuit and district courts of the United States and the Commissioners of the circuit courts may, within their respective jurisdictions, proceed under the law relating to search-warrants, and may issue a search-warrant authorizing and directing the marshal of the United States for the proper district to search for and seize all said counterfeit dies, plates, brands, engravings on wood, stone, metal, or other substance, moulds, and said counterfeit trade-marks, colorable imitations thereof, labels, brands, stamps, wrappers, engravings on paper, or other substance, and said empty boxes, envelopes, wrappers, cases, bottles, or other packages that can be found ; and upon satisfactory proof being made that said counterfeit dies, plates, brands, engravings on wood, stone, metal, or other substance, moulds, counterfeit trade-marks, colorable imitations thereof, labels, brands, stamps, wrappers, engravings on paper or other substance, empty boxes, envelopes, wrappers, cases, bottles, or other packages, are to be used by the holder or owner for the purposes of deception and fraud, that any of said judges shall have full power to order all said counterfeit dies, plates, brands, engravings on wood, stone, metal, or other substance, moulds, counterfeit trade-marks, colorable imitations thereof, labels, brands, stamps, wrappers, engravings on paper or other substance, empty boxes, envelopes, wrappers, cases, bottles, or other packages, to be publicly destroyed.

SEC. 8. That any person who shall, with intent to defraud any person or persons, knowingly and wilfully aid or abet in the violation of any of the provisions of this act, shall, upon conviction thereof, be punished by a fine not exceeding five hundred dollars, or imprisonment not more than one year, or both such fine and imprisonment.

THE
COPYRIGHT LAWS
OF THE
UNITED STATES.

IN FORCE JULY 1, 1891.

FROM THE REVISED STATUTES OF THE UNITED STATES, IN FORCE DECEMBER 1, 1873, AS AMENDED BY ACT APPROVED JUNE 18, 1874, AND FURTHER AMENDED MARCH 3, 1891.

SEC. 4948. All records and other things relating to copyrights and required by law to be preserved shall be under the control of the Librarian of Congress, and kept and preserved in the Library of Congress; and the Librarian of Congress shall have the immediate care and supervision thereof, and, under the supervision of the Joint Committee of Congress on the Library, shall perform all acts and duties required by law touching copyrights.

SEC. 4949. The seal provided for the office of the Librarian of Congress shall be the seal thereof, and by it all records and papers issued from the office, and to be used in evidence, shall be authenticated.

SEC. 4950. The Librarian of Congress shall give a bond, with sureties, to the Treasurer of the United States, in the sum of five thousand dollars, with the condition that he will render to the proper officers of the Treasury a true account of all moneys received by virtue of his office.

SEC. 4951. The Librarian of Congress shall make an annual report to Congress of the number and description of copyright publications for which entries have been made during the year.

SEC. 4952. The author, inventor, designer or proprietor of any book, map, chart, dramatic or musical composition, engraving, cut, print, or photograph or negative thereof, or of a painting, drawing, chromo, statue, statuary, and of models or designs intended to be perfected as works of the fine arts, and the executors, administrators, or assigns of any such person, shall, upon complying with the provisions of this chapter, have the sole liberty of printing, reprinting, publishing, completing, copying, executing, finishing, and vending the same; and, in case of dramatic composition, of publicly performing or representing it or causing it to be performed or represented by others; and authors or their assigns shall have exclusive right to dramatize and translate any of their works for which copyright shall have been obtained under the laws of the United States.

SEC. 4953. Copyrights shall be granted for the term of twenty-eight years from the time of recording the title thereof, in the manner hereinafter directed.

SEC. 4954. The author, inventor, or designer, if he be still living, or his widow or children, if he be dead, shall have the same exclusive right continued for the further term of fourteen years, upon recording the title of the work or description of the article so secured a second time, and complying with all other regulations in regard to original copyrights, within six months before the expiration of the first term; and such persons shall, within two months from the date of said renewal, cause a copy of the record thereof to be published in one or more newspapers printed in the United States for the space of four weeks.

SEC. 4955. Copyrights shall be assignable in law by any instrument of writing, and such assignment shall be recorded in the office of the Librarian of Congress within sixty days after its execution; in default of which it shall be void as against any subsequent purchaser or mortgagee for a valuable consideration, without notice.

SEC. 4956. No person shall be entitled to a copyright unless he shall, on or before the day of publication in this or any foreign country, deliver at the office of the Librarian of Congress, or deposit in the mail within the United States, addressed to the Librarian of Congress, at Washington, District of Columbia, a printed copy of the title of the book, map, chart, dramatic or musical composition, engraving, cut, print, photograph, or chromo, or a description of the painting, drawing, statue, statuary, or a model or design for a work of the fine arts for which he desires a copyright, nor unless he shall also, not later than the day of the publication thereof in this or any foreign country, deliver at the office of the Librarian of Congress, at Washington, District of Columbia, or deposit in the mail within the United States, addressed to the Librarian of Congress, at Washington, District of Columbia, two copies of such copyright book, map, chart, dramatic or musical composition, engraving, chromo, cut, print, or photograph, or in case of a painting, drawing, statue, statuary, model, or a design for a work of the fine arts, a photograph of same; *Provided*, That in the case of a book, photograph, chromo, or lithograph, the two copies of the same required to be delivered or deposited as above shall be printed from type set within the limits of the United States, or from plates made therefrom, or from negatives, or drawings on stone made within the limits of the United States, or from transfers made therefrom. During the existence of such copyright the importation into the United States of any book, chromo, or lithograph, or photograph, so copyrighted, or any edition or editions thereof, or any plates of the same not made from type set, negatives, or drawings on stone made within the limits of the United States, shall be, and it is hereby prohibited, except in the cases specified in paragraphs five hundred and twelve to five hundred and sixteen, inclusive, in section two of the act en-

COPYRIGHT LAWS OF 1891. 95

titled "An act to reduce the revenue and equalize the duties on imports, and for other purposes," approved October first, eighteen hundred and ninety; and except in the case of persons purchasing for use and not for sale, who import subject to the duty thereon, not more than two copies of such book at any one time; and except in the case of newspapers and magazines, not containing in whole or in part matter copyrighted under the provisions of this act, unauthorized by the author, which are hereby exempted from prohibition of importation: *Provided, nevertheless*, 'That in the case of books in foreign languages, of which only translations in English are copyrighted, the prohibition of importation shall apply only to the translation of the same. and the importation of the books in the original language shall be permitted.

SEC. 4957. The Librarian of Congress shall record the name of such copyright book, or other article, forthwith in a book to be kept for that purpose, in the words following: "Library of Congress, to wit: Be it remembered that on the ―― day of ――, ――, A. B., of ――, hath deposited in this office the title of a book (map, chart, or otherwise, as the case may be, or description of the article), the title or description of which is in the following words, to wit: (here insert the title or description), the right whereof he claims as author (originator or proprietor, as the case may be), in conformity with the laws of the United States respecting copyrights. C. D., Librarian of Congress." And he shall give a copy of the title or description, under the seal of the Librarian of Congress, to the proprietor whenever he shall require it.

SEC. 4958. The Librarian of Congress shall receive from the persons to whom the services designated are rendered the following fees.

First. For recording the title or description of any copyright book or other article, fifty cents.

Second. For every copy under seal of such record actually given to the person claiming the copyright, or his assigns, fifty cents.

Third. For recording and certifying any instrument of writing for the assignment of a copyright, one dollar.

Fourth. For every copy of an assignment, one dollar.

All fees so received shall be paid into the Treasury of the United States: *Provided*, That the charge for recording the title or description of any article entered for copyright, the production of a person not a citizen or resident of the United States, shall be one dollar, to be paid as above into the Treasury of the United States, to defray the expenses of lists of copyrighted articles as hereinafter provided for.

And it is hereby made the duty of the Librarian of Congress to furnish to the Secretary of the Treasury copies of the entries of titles of all books and other articles wherein the copyright has been completed by the deposit of two copies of such book printed from type set within the limits of the United States, in acordance with the provisions of this act, and by the deposit of two copies of such other article made or produced in the United States; and the Secretary of the Treasury is hereby directed to prepare and print, at intervals of not more than a week, catalogues of such title-entries for distribution to the collectors of customs of the United States and to the postmasters of all post offices receiving foreign mails, and such weekly lists, as they are issued, shall be furnished to all parties desiring them, at a sum not exceeding five dollars per annum; and the Secretary and the Postmaster-General are hereby empowered and required to make and enforce such rules and regulations as shall prevent the importation into the United States, except upon the conditions above specified, of all articles prohibited by this act.

SEC. 4959. The proprietor of every copyright book or other article shall deliver at the office of the Librarian of Congress, or deposit in the mail, addressed to the Librarian of Congress, at Washington, District of Columbia, a copy of every subsequent edition wherein any substantial changes shall be made: *Provided, however,* That the alteration, revisions, and additions made to books by foreign authors, heretofore published, of which new additions shall appear subsequently to the taking effect of this act, shall be held and deemed capable of being copyrighted as above provided for in this act, unless they form a part of the series in course of publication at the time this act shall take effect.

SEC. 4960. For every failure on the part of the proprietor of any copyright to deliver, or deposit in the mail, either of the published copies, or description, or photograph, required by Sections 4956 and 4959, the proprietor of the copyright shall be liable to a penalty of twenty-five dollars, to be recovered by the Librarian of Congress, in the name of the United States, in an action in the nature of an action of debt, in any district court of the United States within the jurisdiction of which the delinquent may reside or be found.

SEC. 4961. The postmaster to whom such copyright book, title, or other article is delivered shall, if requested, give a receipt therefor; and when so delivered he shall mail it to its destination.

SEC. 4962. No person shall maintain an action for the infringement of his copyright unless he shall give notice thereof by inserting in the several copies of every edition published on the title-page, or the page immediately following, if it be a book; or if a map, chart, musical composition, print, cut, engraving, photograph, painting, drawing, chromo, statue, statuary, or model or design intended to be perfected and completed as a work of the fine arts, by inscribing upon some visible portion thereof, or of the substance on which the same shall be mounted, the following words, viz.: " Entered according to act of Congress, in the year ——, by A. B., in the office of the Librarian of Congress at Washington ;" or, at his option, the word "Copyright," together with the year the copyright was entered, and the name of the party by whom it was taken out, thus: "Copyright, 18—, by A. B."

SEC. 4963. Every person who shall insert or impress such notice, or words of the same purport, in or upon any book, map, chart, dramatic or musical composition, print, cut, engraving, or photograph, or other article, for which he has not obtained a copyright, shall be liable to a penalty of one hundred dollars, recoverable one-half for the person who shall sue for such penalty and one-half to the use of the United States.

SEC. 4964. Every person who, after the recording of the title of any book and the depositing of two copies of such book, as provided by this act, shall, contrary to the provisions of this act, within the term limited, and without the consent of the proprietor of the copyright first obtained in writing, signed in presence of two or more witnesses, print, publish, dramatize, translate, or import, or, knowing the same to be so printed, published, dramatized, translated, or imported, shall sell or expose to sale any copy of such book, shall forfeit every copy thereof to such proprietor, and shall also forfeit and pay such damages as may be recovered in a civil action by such proprietor in any court of competent jurisdiction.

SEC. 4965. If any person, after the recording of the title of any map, chart, dramatic or musical composition, print, cut, engraving, or photograph, or chromo, or of the description of any painting, drawing, statue, statuary, or model or design intended to be perfected and executed as a work of the fine arts, as provided by this act, shall within the term limited, contrary to the provisions of this act, and without the consent of the proprietor of the copyright first obtained in writing, signed in presence of two or more witnesses, engrave, etch, work, copy, print, publish, dramatize, translate, or import, either in whole or in part, or by varying the main design with intent to evade the law, or, knowing the same to be so printed, published, dramatized, translated, or imported, shall sell or expose to sale any copy of such map or other article as aforesaid, he shall forfeit to the proprietor all the plates on which the same shall be copied and every sheet thereof, either copied or printed, and shall further forfeit one dollar for every sheet of the same found in his possession, either printing, printed, copied, published, imported, or exposed for sale, and in case of painting, statue or statuary, he shall forfeit ten dollars for every copy of the same in his possession, or by him sold or exposed for sale; one-half thereof to the proprietor and the other half to the use of the United States.

SEC. 4966. Any person publicly performing or representing any dramatic composition for which a copyright has been obtained, without the consent of the proprietor thereof, or his heirs or assigns, shall be liable for damages therefor; such damages in all cases to be assessed at such sum, not less than one hundred dollars for the first, and fifty dollars for every subsequent performance, as to the court shall appear to be just.

Sec. 4967. Every person who shall print or publish any manuscript whatever without the consent of the author or proprietor first obtained, shall be liable to the author or proprietor for all damages occasioned by such injury.

Sec. 4968. No action shall be maintained in any case of forfeiture or penalty under the copyright laws, unless the same is commenced within two years after the cause of action has arisen.

Sec. 4969. In all actions arising under the laws respecting copyrights the defendant may plead the general issue, and give the special matter in evidence.

Sec. 4970. The circuit courts, and district courts having the jurisdiction of circuit courts, shall have power, upon bill in equity, filed by any party aggrieved, to grant injunctions to prevent the violation of any right secured by the laws respecting copyrights, according to the course and principles of courts of equity on such terms as the court may deem reasonable.

[The foregoing are the Copyright Statutes as amended March 3d, 1891, by the act entitled "An act to amend title sixty, chapter three, of the Revised Statutes of the United States, relating to copyrights." The sections of the law of 1874 which by the act of March 3, 1891, were changed so as to read as above, were sections 4952, 4954, 4956, 4958, 4959, 4963, 4964, 4965, 4967, 4971. By the same act, March 3, 1891, the following additional provisions were adopted:]

Amendments approved March 3, 1891.

Sec. 11. That for the purpose of this act each volume of a book in two or more volumes, when such volumes are published separately, and the first one shall not have been issued before this act shall take effect, and each number of a periodical shall be considered an independent publication, subject to the form of copyrighting as above.

Sec. 12. That this act shall go into effect on the first day of July, Anno Domini eighteen hundred and ninety-one.

Sec. 13. That this act shall only apply to a citizen or subject of a foreign state or nation, when such foreign state or nation permits to citizens of the United States of America the benefit of copyright on substantially the same basis as its own citizens; or when such foreign state or nation is a party to an international agreement which provides for reciprocity in the granting of copyright, by the terms of which agreement the United States of America may, at its pleasure, become a party to such agreement. The existence of either of the conditions aforesaid shall be determined by the President of the United States by proclamation made from time to time as the purposes of this act may require.

Amendments approved June 18, 1874.

Be it enacted by the Senate and House of Representatives of the United States of America in Congress Assembled, That no person shall maintain an action for the infringement of his copyright unless he shall give notice thereof by inserting in the several copies of every edition published, on the title-page or the page immediately following, if it be a book; or if a map, chart, musical composition, print, cut, engraving, photograph, painting, drawing, chromo, statue, statuary, or model or design intended to be perfected and completed as a work of the fine arts, by inscribing upon some visible portion thereof, or of the substance on which the same shall be mounted, the following words, viz.: "Entered according to act of Congress, in the year ——, by A. B., in the office of the Librarian of Congress, at Washington;" or, at his option, the word "Copyright," together with the year the copyright was entered, and the name of the party by whom it was taken out; thus—"Copyright, 18—, by A. B."

SEC. 2. That for recording and certifying any instrument of writing for the assignment of a copyright, the Librarian of Congress shall receive from the persons to whom the service is rendered, one dollar; and for every copy of an assignment, one dollar; said fee to cover, in either case, a certificate of the record, under seal of the Librarian of Congress; and all fees so received shall be paid into the Treasury of the United States.

COPYRIGHTS FOR LABELS.

SEC. 3. That in the construction of this act, the words "engraving," "cut," and "print" shall be applied only to pictorial illustrations or works connected with the fine arts, and no prints or labels designed to be used for any other articles of manufacture shall be entered under the copyright law, but may be registered in the Patent Office. And the Commissioner of Patents is hereby charged with the supervision and control of the entry or registry of such prints or labels, in conformity with the regulations provided by law as to copyright of prints, except that there shall be paid for recording the title of any print or label not a trade-mark, six dollars, which shall cover the expense of furnishing a copy of the record under the seal of the Commissioner of Patents, to the party entering the same.

SEC. 4. That all laws and parts of laws inconsistent with the foregoing provisions be and the same are hereby repealed.

SEC. 5. That this act shall take effect on and after the first day of August, eighteen hundred and seventy-four.

Repeal Provisions.

TITLE LXXIV., Rev. Stat., p. 1091:

SEC. 5595. The foregoing seventy-three titles embrace the

statutes of the United States general and permanent in their nature, in force on the 1st day of December, one thousand eight hundred and seventy-three, as revised and consolidated by commissioners appointed under an act of Congress, and the same shall be designated and cited, as The Revised Statutes of the United States.

SEC. 5596. All acts of Congress passed prior to said first day of December, one thousand eight hundred and seventy-three, any portion of which is embraced in any section of said revision, are hereby repealed, and the section applicable thereto shall be in force in lieu thereof; all parts of such acts not contained in such revision, having been repealed or superseded by subsequent acts, or not being general and permanent in their nature: *Provided*, That the incorporation into said revision of any general and permanent provision, taken from an act making appropriations, or from an act containing other provisions of a private, local, or temporary character, shall not repeal, or in any way affect any appropriation, or any provision of a private, local, or temporary character, contained in any of said acts, but the same shall remain in force; and all acts of Congress passed prior to said last-named day, no part of which are embraced in said revision, shall not be affected or changed by its enactment.

SEC. 5597. The repeal of the several acts embraced in said revision shall not affect any act done, or any right accruing or accrued, or any suit or proceeding had or commenced in any civil cause before the said repeal, but all rights and liabilities under said acts shall continue, and may be enforced in the same manner, as if said repeal had not been made; nor shall said repeal in any manner affect the right to any office, or change the term or tenure thereof.

SEC. 5598. All offences committed, and all penalties or forfeitures incurred under any statute embraced in said revision prior to said repeal, may be prosecuted and punished in the same manner and with the same effect as if said repeal had not been made.

SEC. 5599. All acts of limitation, whether applicable to civil causes and proceedings, or to the prosecution of offences, or for the recovery of penalties or forfeitures, embraced in said revision and covered by said repeal, shall not be affected thereby, but all suits, proceedings, or prosecutions, whether civil or criminal, for causes arising or acts done or committed prior to said repeal, may be commenced and prosecuted within the same time as if said repeal had not been made.

COPYRIGHT LAWS OF 1891.

SEC. 5600. The arrangement and classification of the several sections of the revision have been made for the purpose of a more convenient and orderly arrangement of the same, and, therefore, no inference or presumption of a legislative construction is to be drawn by reason of the title under which any particular section is placed.

SEC. 5601. The enactment of the said revision is not to affect or repeal any act of Congress passed since the 1st day of December, one thousand eight hundred and seventy-three, and all acts passed since that date are to have full effect as if passed after the enactment of this revision, and so far as such acts vary from, or conflict with, any provision contained in said revision, they are to have effect as subsequent statutes, and as repealing any portion of the revision inconsistent therewith.

ELECTRIC BATTERIES AND MAGNETS.

THE following directions, if carefully followed, will enable any person to make a working galvanic battery, electro-magnet, and a needle telegraph, all at a cost of not exceeding ten or twenty cents:

Take a bit of common sheet zinc—stove zinc—six inches long, four inches wide. Bend it up two inches from the end, nick the bent end, and wind the naked end of a copper wire around the zinc at the nick. Use copper wire, size of a common pin—insulated wire, that is, copper wire covered with thin paper or cotton, or paint or varnish. Leave the ends of the wires uncovered. Cover the zinc with common white printing paper, two thicknesses, as shown at 1. The white represents the paper. Provide a bit of thin sheet lead, same size as the zinc, bent in and nicked, but not covered with paper. Attach one end of a copper wire, as shown at 2. Provide a common tea saucer, in which place one ounce of sulphate of copper (to be had for a few cents at any drug store); pour on warm water; fill saucer two-thirds full; let stand until dissolved. Now put the zinc (1) in saucer, and put the lead (2) on top of the zinc, the ends standing above the liquid as shown. You now have a com-

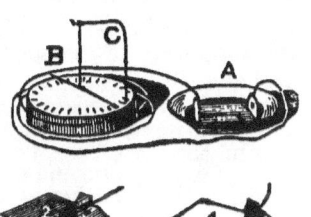

plete and tolerably strong galvanic battery. When the free ends of the two wires are touched together, a spark will be seen at the moment of junction. The circuit is made or closed by placing the two wires together. The circuit is opened or broken by separating the wires.

An electro-magnet may be made by providing a small wrought iron staple, and grinding off the pointed ends. Wind both legs with the insulated copper wire. The mode of winding and carrying the wire from one leg to the other is shown in the cut. Both legs should, when finished, be closely wound, as shown on the left side.

Now connect the end of the wire of one of the legs with one of the battery wires, and the wire of the other leg with the other battery wire, and you will find that the staple is magnetized. The magnetism ceases when the wires are separated. Put a steel knife blade on one leg of the magnet, and the knife becomes permanently magnetized. A needle drawn across the magnet is also magnetized, and if suspended by a thread it becomes a compass, and will point north and south. To make such a compass, magnetize a sewing needle; suspend it by a silk fiber drawn from a piece of sewing silk; attach the silk to a bit of bent wire, C, the bottom of which is stuck into, and supported by, a piece of round cork, as shown in the first cut.

Place the compass thus made on a table, and the needle will point north and south. Now place loosely on the table around the cork base, two coils or turns of the insulated wire. Connect one end of the wire with one of the battery wires. Join the other end of the coil wire to the other battery wire. An electrical current will now go through the coil, which will move the needle. This is the needle telegraph. By alternately separating and joining the wire of the battery and the coil wire, the needle swings. One swing may be called A; two swings, B; three, C, etc. In this way intelligible signals are sent. A needle smaller than that here shown in the cut is used for telegraphing through the cable under the Atlantic Ocean between Europe and America. On land lines in England larger needles are used, but in this country a magnet is used which draws down a lever with such force as to make a click. A long click and a short click mean A; a short click, E; two short clicks, I. The battery wire and the coil wire are alternately joined and separated by a pivoted finger lever called a key.

DISTINGUISHED AMERICAN INVENTORS.

BENJAMIN FRANKLIN; b. Boston, 1706; d. 1790; at 12, printer's apprentice, fond of useful reading; 27 to 40, teaches himself Latin, etc., makes various useful improvements; at 40, studies electricity; 1752, brings electricity from clouds by kite, and invents the lightning rod.

ELI WHITNEY, inventor of the cotton-gin; b. Westborough, Mass., 1765; d. 1825: went to Georgia 1792, as teacher; 1793, invents the cotton-gin, prior to which a full day's work of one person was to clean by hand one pound of cotton; one machine performs the labor of five thousand persons; 1800, founds Whitneyville, makes firearms by the interchangeable system for the parts.

ROBERT FULTON; b. Little Britain, Pa., 1765; d. 1825; artist painter; invents steamboat 1793; invents submarine torpedoes 1797 to 1801; builds steamboat in France 1803; launches passenger boat Clermont at N. Y. 1807, and steams to Albany; 1812, builds steam ferryboats; 1814, builds first steam war vessel.

THOMAS BLANCHARD; b. 1788, Sutton, Mass.; d. 1864; invented tack machine 1806; builds successful steam carriage 1825; builds the stern-wheel boat for shallow waters, now in common use on Western rivers; 1843, patents the lathe for turning irregular forms, now in common use all over the world for turning lasts, spokes, axe-handles, gun-stocks, hat-blocks, tackle-blocks, etc.

CYRUS H. MCCORMICK, inventor of harvesting machines; b. Walnut Grove, Va., 1809; in 1851 he exhibited his invention at the World's Fair, London, with practical success. The mowing of one acre was one man's day's work; a boy with a mowing-machine now cuts ten acres a day. Mr. McCormick's patents made him a millionaire.

CHARLES GOODYEAR, inventor and patentee of the simple mixture of rubber and sulphur, the basis of the present great rubber industries throughout the world; b. New Haven, Conn., 1800; in 1839, by the accidental mixture of a bit of rubber and sulphur on a red-hot stove, he discovered the process of vulcanization. The Goodyear patents proved immensely profitable.

DISTINGUISHED AMERICAN INVENTORS.

CELEBRATED AMERICAN INVENTIONS. 105

DISTINGUISHED AMERICAN INVENTORS.

ELIAS HOWE, inventor of the modern sewing machine; b. Spencer, Mass., 1819; d. 1867; machinist; sewing machine patented 1846. From that time to 1854 his priority was contested, and he suffered from poverty, when a decision of the courts in his favor brought him large royalties, and he realized several millions from his patent.

SAMUEL F. B. MORSE, inventor and patentee of electric telegraph; b. Charlestown, Mass., 1791; d. 1872; artist painter; exhibited first drawings of telegraph 1832; half-mile wire in operation 1835; caveat 1837; Congress appropriated $30,000, and in 1844 first telegraph line from Washington to Baltimore was opened; after long contests, the courts sustained his patents, and he realized from them a large fortune.

JAMES B. EADS; b. 1820; author and constructor of the great steel bridge over the Mississippi at St. Louis, 1867, and the jetties below New Orleans 1876. His remarkable energy was shown in 1861, when he built and delivered complete to Government, all within sixty-five days, seven iron-plated steamers, 600 tons each; subsequently other steamers. Some of the most brilliant successes of the Union arms were due to his extraordinary rapidity in constructing these vessels.

JOHN ERICSSON; b. in Sweden, 1803; d. in New York, 1889. Designed the locomotive, and competed with Stephenson, London. Designed the screw propeller and mechanism for ships and tugs. 1839 came to the United States, and produced many important works. Designed war ship Princeton; constructed heavy ordnance; invented hot-air engine; built the first iron-clad monitor in 1862; designed torpedo boats; made solar engine, etc.

THOMAS ALVA EDISON; b. in Ohio, 1847. Printer's boy, then telegraph operator; 1864 invented quadruplex telegraph. Many electrical inventions followed, including the telephone, the incandescent system of lighting, the dynamo-magnetic separator for metals, etc. In 1878 he produced the phonograph. Has patented over six hundred inventions, bringing him great wealth. He is justly regarded as the greatest intellect of modern times.

ALEXANDER GRAHAM BELL, inventor of the telephone; b. in Scotland, 1847. In 1876 produced the electric telephone; author of many valuable scientific contributions. The telephone brought him immense wealth, and to his associates far more than to himself.

MECHANICAL MOVEMENTS.

IN the construction of models, or machinery, the skilful mechanic and inventor will study to avoid clumsiness in the arrangement of parts, and will naturally take pride in selecting, as far as possible, the simplest and best forms of mechanical movements. As suggestive for this purpose we have brought together and condensed an extensive series of mechanical movements. Here the mechanic may find at a glance the movement suited for his purpose, and may see the separate parts best adapted to any special combination of mechanism.

The following is a brief description of the various movements, as numbered:

1. Shaft coupling. 2. Claw coupling. 3, 4. Lever couplings. On the driving shaft, a disk with spurs is mounted, and to the shaft to be driven a lever is hinged. By causing this lever to catch in the spurs of the disk, the coupling is effected. 5. Knee or rose coupling, of which 26 is a side view. 6. Universal joint. 7, 8. Disk and spur coupling. 9. Prong and spur lever coupling.

10. Fast and loose pulley. 11. Sliding gear, the journal boxes of one of the wheels being movable. 12. Friction clutch. By tightening or releasing a steel band, encircling a pulley on the shaft, the machinery is thrown in or out of gear. 13, 14. Shoe and lever brakes. 15, 16. Change of motion by sheaves. 17. Spiral flanged shaft. 18. Connected with the rod are pawl links, catching into ratchet-teeth in the wheel to which rotary motion is to be imparted. When the rod moves in one direction, one of the pawls acts; and when the rod moves in the opposite direction, the other pawl acts in the same direction as the first. 19. The reciprocating motion of a rod is converted into rotary motion of the fly-wheel by a weight suspended from a cord, which passes over a small pulley that connects with a treadle, from which the motion is transmitted to the fly-wheel.

20. "Flying horse," used in fairs for amusement. By pulling the cords radiating from the crank, the persons occupying the seats or horses on the ends of the arms are enabled to keep the apparatus in motion. 21, 22. Bow-string arrangements, to convert reciprocating into rotary motion. 23. Same purpose by differential screw. 24. The same by double rack and wheels. 25. Coupling for square shafts. 26. Side view of Fig. 5. 27. Sliding-spur pulley coupling. 28. Lever with bearing roller to tighten pulley bands. 29. Chain wheel.

30. Reciprocating rectilinear into reciprocating rotary

motion by two racks and cog-wheel. 31. Oblique-toothed wheels. 32. Worm and worm-wheel. 33, 34. Claw coupling with hinged lever. 35, 36. Disk couplings, with lugs and cavities. 37. Disk coupling with screw bolts. 38, 39, 40. Shaft couplings.
41. Face view of Fig. 12. 42. Friction cones. 43. Friction pulleys. 44. Self-releasing coupling. Disks with oblique teeth. If the resistance to the driven shaft increases beyond a certain point, the disks separate. 45. Hoisting blocks. 46. Elbow crank, for changing motion. 47. Reciprocating into rotary motion by zigzag groove on cylinder. 48. Another form of Fig. 29. 49. Reciprocating into a rotary motion.
50. Same purpose. 51. Same purpose, by double rack and two ratchet pinions. When the double rack moves in one direction, one pinion is rigid with the shaft; when the rack moves in the opposite direction, the other pinion is rigid, and a continuous rotary motion is imparted to the fly-wheel shaft. 52. Reciprocating into oscillating. 53. Rotary into reciprocating. By the action of the wheel-pins the carriage is moved in one direction, and by the action of said pins on an elbow-lever, it is moved in the opposite direction. 54. Stamp rod and lifting cam. 55. For giving reciprocating motion to rack. 56. Same motion to a bar with slot, by means of an eccentric pin projecting from a revolving disk, and catching in the slot. 57. Walking-beam and fly-wheel. 58. Reciprocating motion to pump or other rod by means of eccentric disk and friction rollers. See 81 and 104. 59. Hoisting crane.
60. Friction gears. See 43. 61. Rotary into reciprocating by rising and falling pinion acting on endless rack. 62. By the revolving cam, a rising and falling or a reciprocating rectilinear motion is imparted to a drum. 63. Reciprocating motion to a frame by means of endless rack and pinion. 64. Reciprocating rectilinear motion to a toothed rack by a toothed segment on a lever-arm, which is subjected to the action of a weight, and of an eccentric wrist-pin, projecting from a revolving disk. 65. Reciprocating motion to a rod. The wheels are of different diameters, and consequently the rod has to rise and fall as the wheels revolve. (See 110.) 66. Cam and elbow lever. 67. Rod reciprocates by means of cam. 68. Revolving into reciprocating motion, by an endless segmental rack and pinion, the axle of which revolves and slides in a slot toward and from the rack. This rack is secured to a disk, and a rope round said disk extends to the body to which a reciprocating motion is to be imparted. 69. Elliptic gears.

MECHANICAL MOVEMENTS.

70. Bevel gear. 71. Worm and worm wheel. 72. Transmitting motion from one axle to another, with three different velocities, by means of toothed segments of unequal diameters. 73. Continuous revolving into reciprocating, by a cam-disk acting on an oscillating lever. 74. Intermittent revolving motion to a shaft with two pinions, and segment gear-wheel on end of shaft. 75. Oscillating lever, carrying pawls which engage teeth in the edges of a bar to which rectilinear motion is imparted. 76. Oscillating lever, connects by a link with a rod to which a rectilinear motion is imparted. 77. Oscillating lever and pawls, which gear in the ratchet-wheel. 78. Common treadle. 79. Describing on a revolving cylinder a spiral line of a certain given pitch which depends upon the comparative sizes of the pinion and bevel-wheels.
80. Marking a spiral line, the graver moved by a screw. 81. (See Fig. 58.) 82. Plunger and rods. 83. Crosshead and rods. 84. Reciprocating rod guided by friction rollers. 85. Revolving into reciprocating motion, by means of roller-arms, extending from a revolving shaft, and acting on lugs projecting from a reciprocating frame. 86. Crank motion. 87. Reciprocating motion by toothed wheel and spring bar. 88. The shaft carries a taper, which catches against a hook hinged to the drum, so as to carry said drum along and raise the weight on the rope. When the tappet has reached its highest position, the hook strikes a pin, the hook disengages from the tappet, and the weight drops. 89. Reciprocating motion to a rod by means of a groove in an oblique ring secured to a revolving shaft.
90. Double crank. 91. Cam groove in a drum, to produce reciprocating motion. 92. Belts and pulleys. 93. Pulleys, belts, and internal gear. 94. As the rod moves up and down, the teeth of the cog-wheel come in contact with a pawl, and an intermittent rotary motion is imparted to said wheel. 95. By turning the horizontal axles with different velocities, the middle wheel is caused to revolve with the mean velocity. 96. Oscillating lever and cam groove in a disk. 97. Lazy tongs. 98. Oscillating segment and belt over pulleys. 99. Converting oscillating into a reciprocating motion by a cam-slot in the end of the oscillating lever which catches over a pin projecting from one of the sides of a parallelogram which is connected to the rod to which reciprocating motion is imparted.
100. Oscillating motion of a beam into rotary motion. 101. Motion of a treadle into rotary motion. 102. Double-acting beam. 103. Single-acting beam. 104. (See Figures 58 and 81.) 105. Device to steady a piston by a slotted guide-piece, operated by an eccentric on the driving-shaft.

106. Rod operated by two toothed segments. **107.** Two cog-wheels of equal diameter, provided with a crank of the same length, and connected by links with a cross-bar to which the piston-rod is secured. **108.** Device for a rectilinear motion of a piston-rod based on the hypocycloidal motion of a pinion in a stationary wheel with internal gear. If the diameter of the pinion is exactly equal to one half the diameter of the internal gear, the hypocycloid becomes a eight line. **109.** Same purpose as 56.
110. Action similar to 65. **111.** Revolving motion by a circular sliding pinion gearing in an elliptical cog-wheel. **112.** Similar to 96. **113.** Carpenter's clamp. The jaws turn on their pivot-screws, and clamp the board. **114.** An irregular vibratory motion is given to the arm carrying the wheel A, by the rotation of the pinion B. **115.** Intermittent rotary motion of the pinion-shaft, by the continuous rotary motion of the large wheel. The part of the pinion shown next the wheel is cut on the same curve as the plain portion of the circumference, and, therefore, serves as a lock whilst the wheel makes a part of a revolution, and until the pin upon the wheel strikes the guide-piece upon the pinion, when the pinion-shaft commences another revolution. **116.** Stop-motion used in watches to limit the number of revolutions in winding up. The convex curved part, a, b, of the wheel B, serving as the stop. **117.** Several wheels, by connecting-rods, driven from one pulley. **118.** Intermittent circular motion is imparted to the toothed wheel by vibrating the arm B. When the arm B is lifted, the pawl is raised from between the teeth of the wheel, and travelling backward over the circumference again, drops between two teeth on lowering the arm, and draws with it the wheel. **119.** Reciprocating rectilinear motion is given to the bar by the continuous motion of the cam. The cam is of equal diameter in every direction measured across its centre.
120. Mechanism for revolving the cylinder in Colt's fire-arms. When the hammer is drawn back the dog, a, attached to the tumbler, acts on the ratchet, b, on the back of the cylinder, and is held up to the ratchet by a spring, c. **121.** Alternate increasing and diminishing motion, by means of eccentric toothed wheel and toothed cylinder. **122.** Oscillating or pendulum engine. The cylinder swings between trunnions like a pendulum. The piston-rod connects directly with crank. **123.** Intermittent rotary motion. The small wheel is driven, and the friction rollers on its studs move the larger wheel by working against the faces of oblique grooves or projections across the face thereof. **124.** Longitudinal and rotary motion of the rod is produced by its arrangement

between two rotating rollers, the axles of which are oblique to each other. 125. Friction indicator of Roberts. Upon the periphery of the belt-pulley a loaded carriage is placed, its tongue connected with an indicator. With a given load the indicating pointer remains in a given position, no matter what velocity is imparted to the pulley. When the load is changed the indicator changes, thus proving that the friction of wheels is in proportion to load, not velocity. 126. Circular intermittent rectilinear reciprocating motion. Used on sewing-machines for driving the shuttle ; also on three-revolution cylinder printing-presses. 127. Continuous circular into intermittent circular motion. The cam is the driver. 128. Sewing-machine, four-motion feed. The bar B carries the feeding-points or spurs, and is pivoted to slide A. B is lifted by a radial projection on cam C, which at the same time also carries A and B forward. A spring produces the return stroke, and the bar, B, drops by gravity. 129. Patent crank motion, to obviate dead centres. Pressure on the treadle moves the slotted slide A forward until the wrist passes the centre, when the spring B forces the slide against the stops until next forward movement.

130. Four-way cock. 131. One stroke of the piston gives a complete revolution to the crank. 132. Rectilinear motion of variable velocity is given to the vertical bar by rotation of the shaft of the curved arm. 133. Pantagraph for copying, enlarging, and reducing plans, etc. C, fixed point. B, ivory tracing point. A, pencil trace, the lines to be copied with, and B, the pencil, will reproduce it double size. Shift the slide to which C is attached, also the pencil slide, and size of the copy will be varied. 134. Ball-and-socket joint for tubing. 135. Numerical registering device. The teeth of the worm shaft-gear with a pair of worm-wheels of equal diameter, one having one tooth more than the other. If the first wheel has 100 teeth and the second 101, the pointers will indicate respectively 101 and 10.100 revolutions. 136. Montgolfier's hydraulic ram. The right hand valve being kept open by a weight or spring, the current flowing through the pipe in the direction of the arrow, escapes thereby. When the pressure of the water current overcomes the weight of the right valve, the momentum of the water opens the other valve, and the water passes into the. air-chamber. On equilibrium taking place, the left valve shuts and the right valve opens. By this alternate action of the valves, water is raised into the air-chamber at every stroke. 137. Rotary engine. Shaft B and hub C are arranged eccentric to the case. Sliding radial pistons, a, a, move in and out of hub, C. The pistons slide through rolling packings in the hub C.

138. Quadrant engine. Two single-acting pistons, B, B, connect with crank D. Steam is admitted to act on the outer sides of the pistons alternately through valve *a*, and the exhaust is between the pistons. 139. Circular into rectilinear motion. The scolloped wheel communicates motion to the horizontal oscillating rod, and imparts rectilinear movement to the upright bar. 140. Rotary motion transmitted by rolling contact between two obliquely arranged shafts.

MULTUM IN PARVO.

WE have some queer correspondents: One writes to know if we will not be so good as to send a messenger to an address which he gives, up town—distance two and a half miles from our office—to make certain inquiries for him. It would require one and a half hours' time to do the errand, and not a stamp inclosed. Another wants us to write a letter and tell him where to get a combined thermometer and barometer. Another, "Will you be good enough to give me the names and addresses of several of the makers of the best brick machines?" another wants water-wheels; another threshing machines: each writer desires our written opinion as to which is the best device, with our reasons, and not one is thoughtful enough to inclose a fee, or to reflect that to answer his request will consume considerable of our time. Another party wishes us to write to him the recipe for making ornaments out of coal tar, where he can buy the mixture ready for use, and how much chequer-men will sell for in the New York market. For this information he sends us the generous sum of three cents in postage stamps. Mr. C wants us to tell him of some valuable invention, of which he can buy the patent cheap, that would be suitable for him to take to sell, on his travels out West, by towns, counties, etc., three cents inclosed. Others want us to put them in communication with some person who will purchase an interest in their inventions, or manufacture for them, or furnish this or that personal information, our reply to be printed in THE SCIENTIFIC AMERICAN. We are at all times happy to serve our correspondents, but if replies to purely personal errands are expected, a small fee, say from one to five dollars, should be sent.

HARNESS BLACKING.—Melt 1 pound bees-wax, stir in 4 ounces ivory-black, 2 ounces spirits turpentine, 2 ounces Prussian blue ground in oil, and ½ ounce copal varnish. Make into balls. With a brush apply it to harness, and polish with silk gently.

MECHANICAL MOVEMENTS. 113

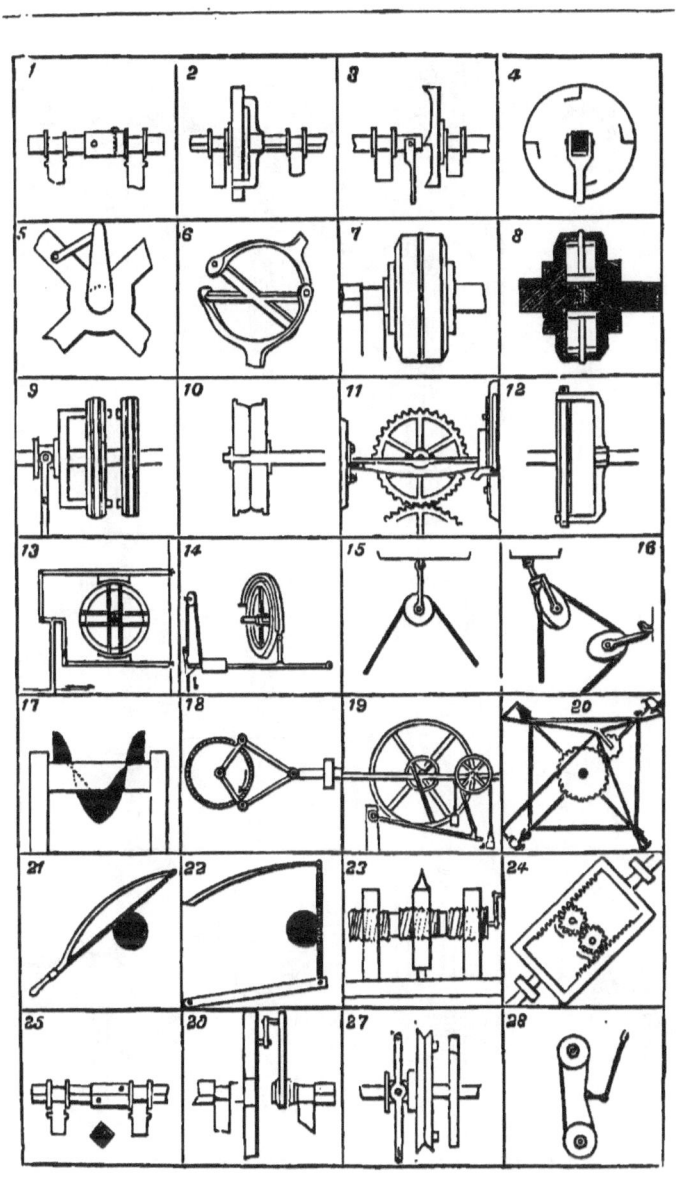

114 MECHANICAL MOVEMENTS.

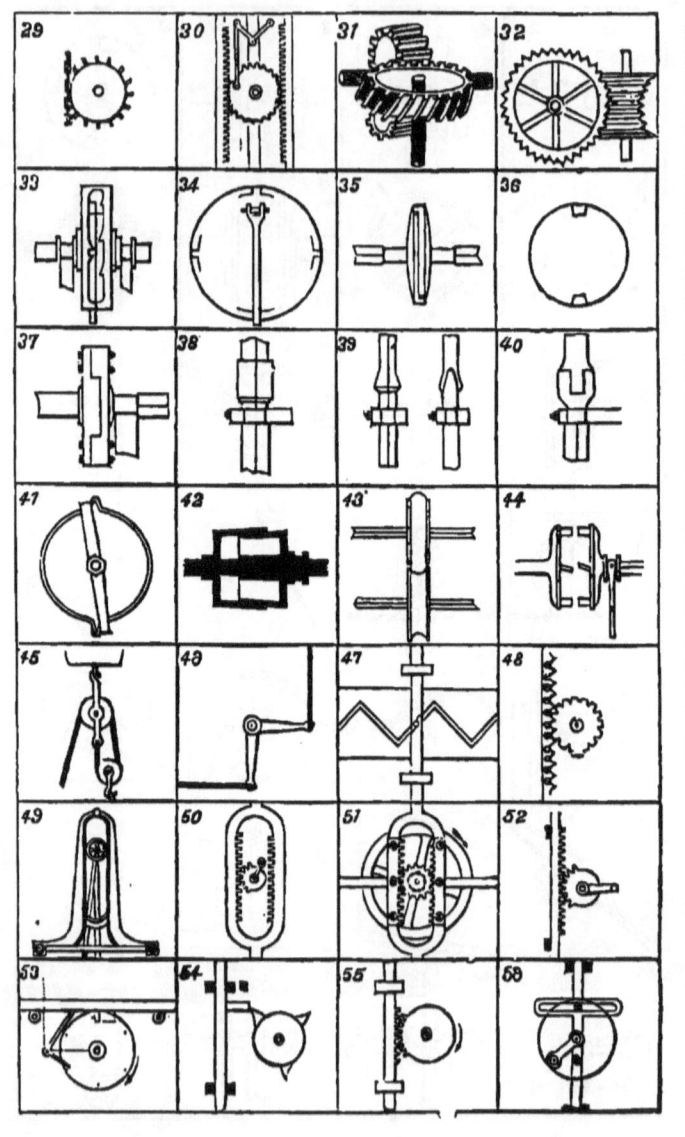

MECHANICAL MOVEMENTS. 115

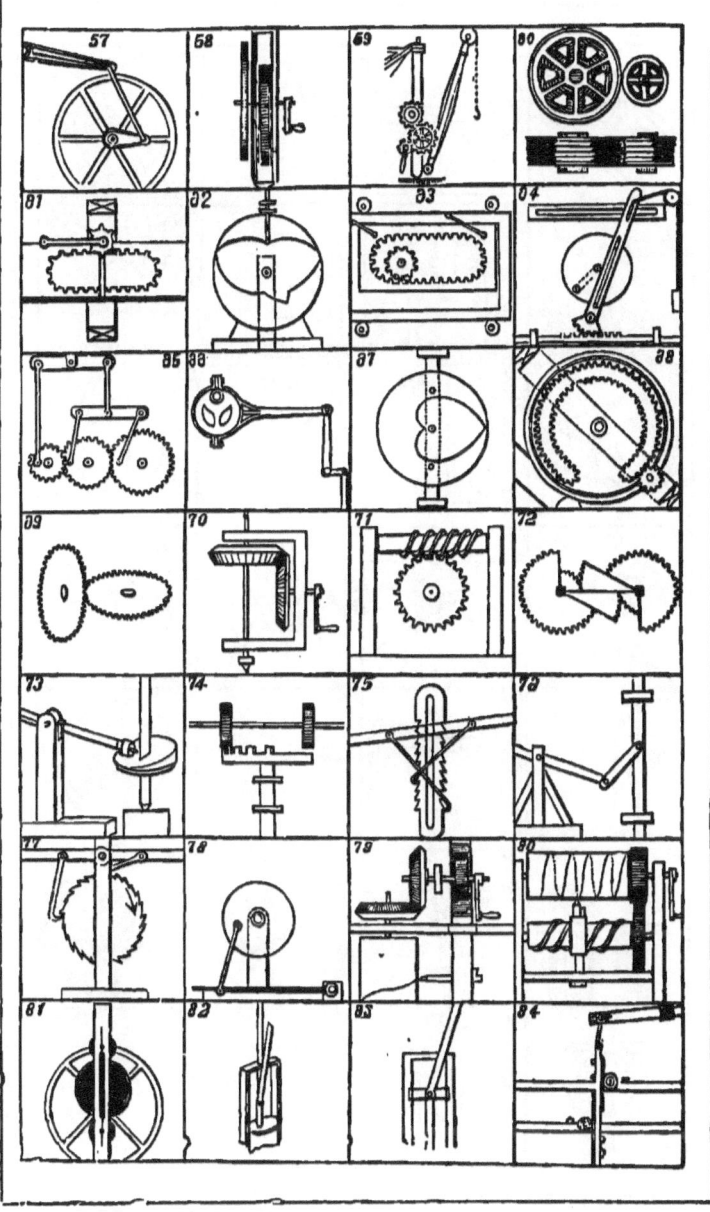

MECHANICAL MOVEMENTS.

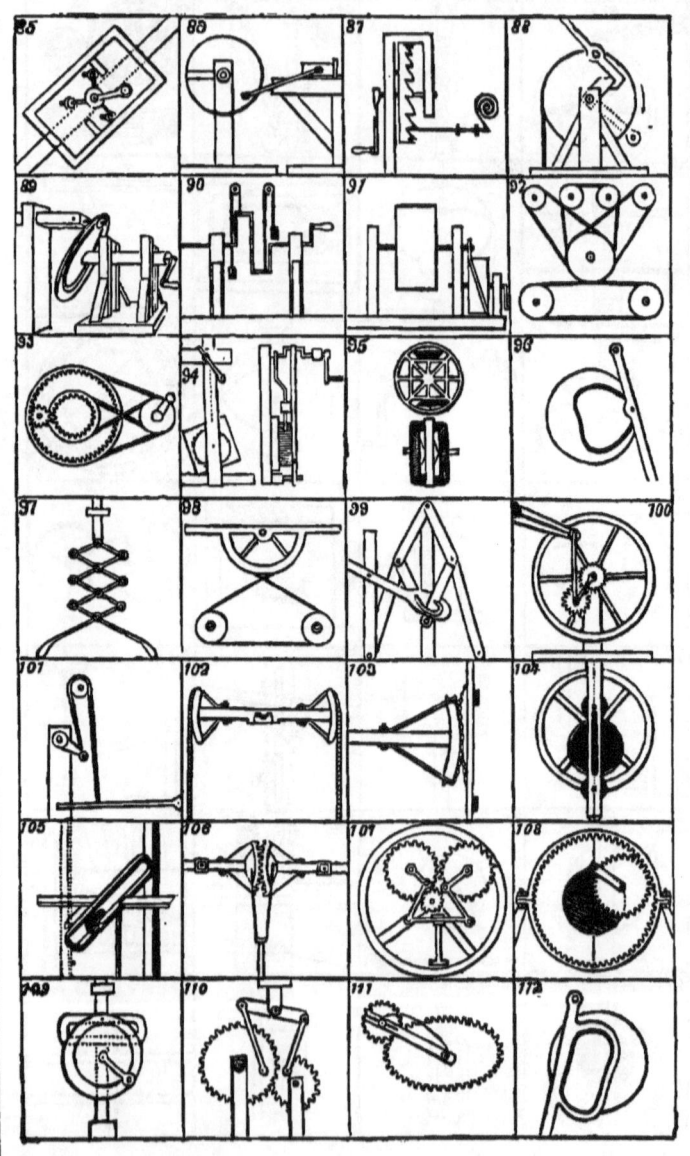

MECHANICAL MOVEMENTS. 117

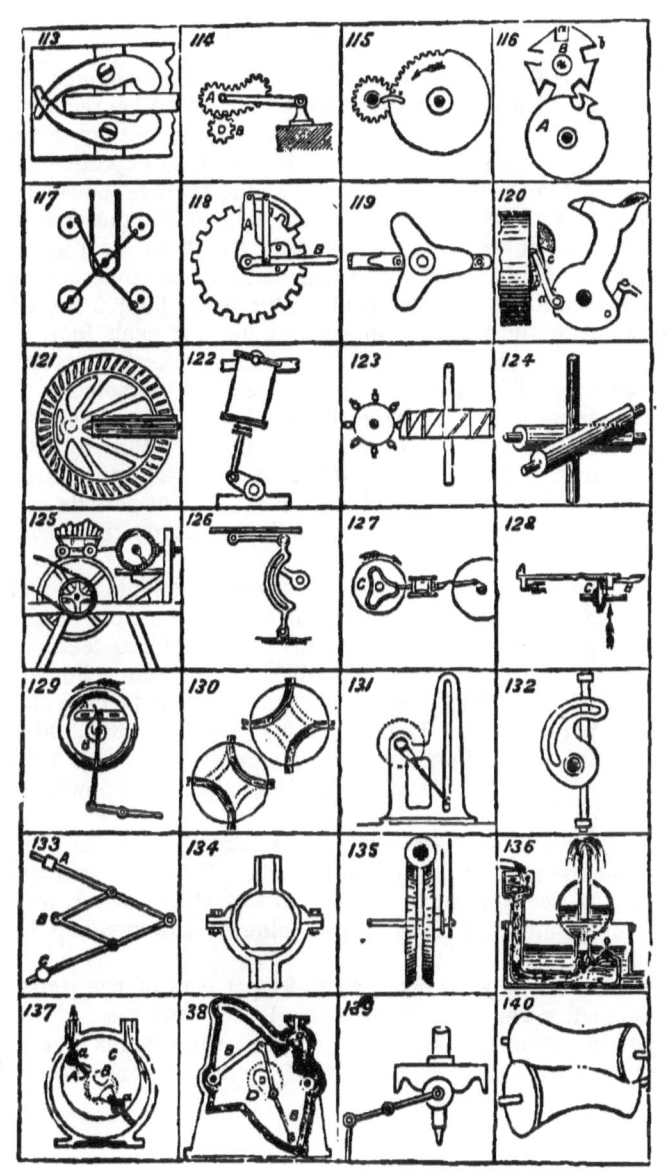

HORSE-POWER.

When Watt began to introduce his steam-engines he wished to be able to state their power as compared with that of horses, which were then generally employed for driving mills. He accordingly made a series of experiments, which led him to the conclusion that the average power of a horse was sufficient to raise about 33,000 lbs. one foot in vertical height per minute, and this has been adopted in England and this country as the general measure of power.

A waterfall has one horse-power for every 33,000 lbs. of water flowing in the stream per minute, for each foot of fall. To compute the power of a stream, therefore, multiply the area of its cross section in feet by the velocity in feet per minute, and we have the number of cubic feet flowing along the stream per minute. Multiply this by $62\frac{1}{2}$, the number of pounds in a cubic foot of water, and this by the vertical fall in feet, and we have the foot-pounds per minute of the fall; dividing by 33,000 gives us the horse-power.

For example: A stream flows through a flume 10 feet wide, and the depth of the water is 4 feet; the area of the cross section will be 40 feet. The velocity is 150 feet per minute—$40 \times 150 = 6000 =$ the cubic feet of water flowing per minute. $6000 \times 62\frac{1}{2} = 375,000 =$ the pounds of water flowing per minute. The fall is 10 feet; $10 \times 375,000 = 3,750,000 =$ the foot-pounds of the water-fall. Divide 3,750,000 by 33,000, and we have $113\frac{21}{33}$ as the horse-power of the fall.

The power of a steam-engine is calculated by multiplying together the area of the piston in inches, the mean pressure in pounds per square inch, the length of the stroke in feet, and the number of strokes per minute; and dividing by 33,000.

Water-wheels yield from 50 to 91 per cent of the water. The actual power of a steam-engine is less than the indicated power, owing to a loss from friction; the amount of this loss varies with the arrangement of the engine and the perfection of the workmanship.

PROPERTIES OF CHARCOAL.

ALTHOUGH charcoal is so combustible, it is, in some respects a very unchangeable substance, resisting the action of a great variety of other substances upon it. Hence posts are often charred before being put into the ground. Grain has been found in the excavations at Herculaneum, which was charred at the time of the destruction of that city, eighteen hundred years ago, and yet the shape is perfectly preserved, so that you can distinguish between the different kinds of grain. While charcoal is itself so unchangeable, it preserves other substances from change. Hence meat and vegetables are packed in charcoal for long voyages, and the water is kept in casks which are charred on the inside. Tainted meat can be made sweet by being covered with it. Foul and stagnant water can be deprived of its bad taste by being filtered through it. Charcoal is a great decolorizer. Ale and porter filtered through it are deprived of their color, and sugar-refiners decolorize their brown syrups by means of charcoal, and thus make white sugar. Animal charcoal, or bone-black, is the best for such purposes, although only one-tenth of it is really charcoal, the other nine-tenths being the mineral portion of the bone.

Charcoal will absorb, of some gases, from eighty to ninety times its own bulk. As every point of its surface is a point of attraction, it is supposed to account for the enormous accumulation of gases in the spaces of the charcoal. But this accounts for it only in part. There must be some peculiar power in the charcoal to change, in some way, the condition of a gas of which it absorbs ninety times its own bulk.—*Hooker.*

SUBSTITUTE FOR THE CRANK.

VARIOUS devices supposed to have advantages over the common crank, have been invented. Our diagram shows one of these forms, which has been re-invented many times, by different inventors. A grooved wheel is employed, and in the groove are two slides, attached respectively, by pivots, to the connecting rod of a piston rod. The reciprocating movement of the piston rod acting upon the connecting rod, causes the rotation of the wheel.

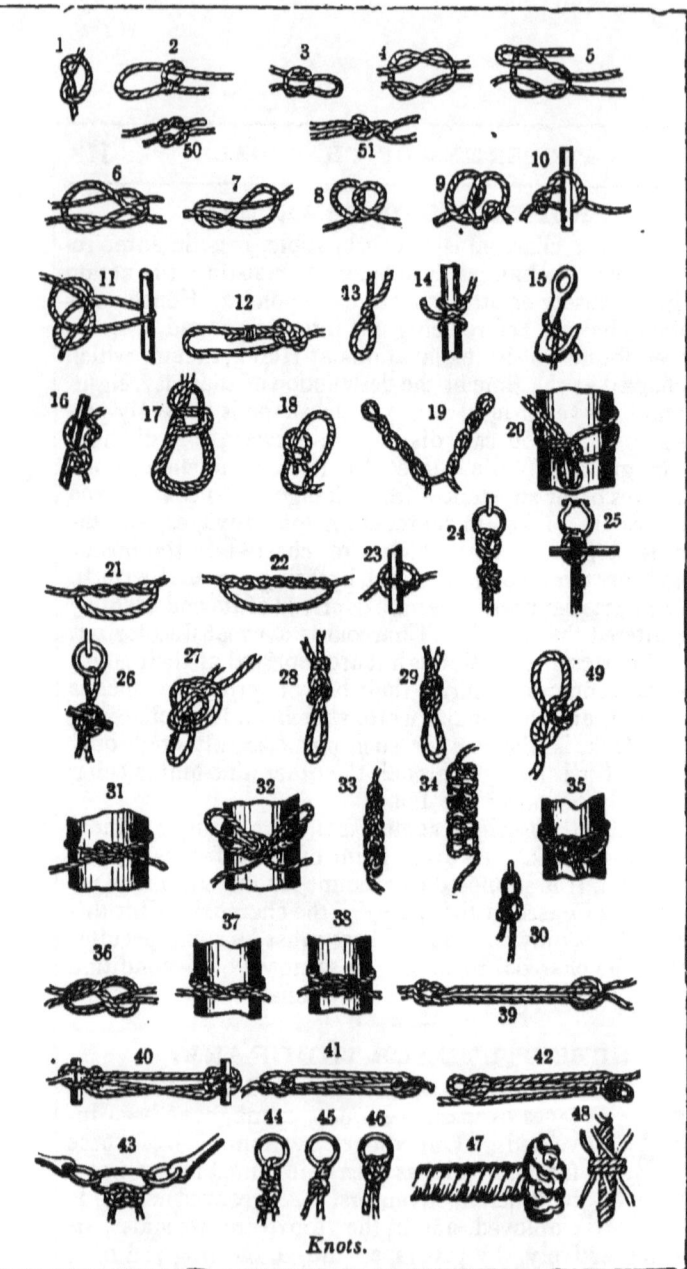

Knots.

KNOTS.

The knots represented on the preceding page of engravings are as follows:

1. Simple overhand knot.
2. Slip-knot seized.
3. Single bow-knot.
4. Square or ruf-knot.
5. Square bow-knot.
6. Weaver's knot.
7. German or figure-of-8 knot.
8. Two half-hitches, or artificer's knot.
9. Double artificer's knot.
10. Simple galley-knot.
11. Capstan or prolonged knot.
12. Bowline-knot.
13. Rolling-hitch.
14. Clove-hitch.
15. Blackwall-hitch.
16. Timber-hitch.
17. Bowline on a bight.
18. Running bowline.
19. Catspaw.
20. Doubled running-knot.
21. Double knot.
22. Six-fold knot.
23. Boat-knot.
24. Lark's head.
25. Lark's head.
26. Simple boat-knot.
27. Loop-knot.
28. Double Flemish knot.
29. Running-knot checked
30. Crossed running-knot.
31. Lashing-knot.
32. Rosette.
33. Chain-knot.
34. Double chain-knot.
35. Double running-knot, with check-knot.
36. Double twist-knot.
37. Builders' knot.
38. Double Flemish knot.
39. English knot.
40. Shortening-knot.
41. Shortening-knot.
42. Sheep-shank.
43. Dog-shank.
44. Mooring-knot.
45. Mooring-knot.
46. Mooring-knot.
47. Pigtail worked on the end of a rope.
48. Shroud-knot.
49. A bend or knot used by sailors in making fast to a spar or a bucket-handle before casting overboard; it will not run. Also used by horsemen for a loop around the jaw of a colt in breaking: the running end, after passing over the head of the animal and through the loop, will not jam therein.
50. A granny's knot.
51. A weaver's knot.

The principle of a knot is, that no two parts which would move in the same direction if the rope were to slip, should lie alongside of and touching each other.

MEASURES OF LENGTH.

MEASURES OF LENGTH.—The subjoined engraving shows at the left a four-inch section of a common rule, the inch divisions being subdivided into twenty-fifths, twelfths, eighths, and tenths. On the right is the French measure, indicating millimetres and centimetres. The French metre is intended to be the one ten-millionth part of the distance from either pole of the earth to the equator.

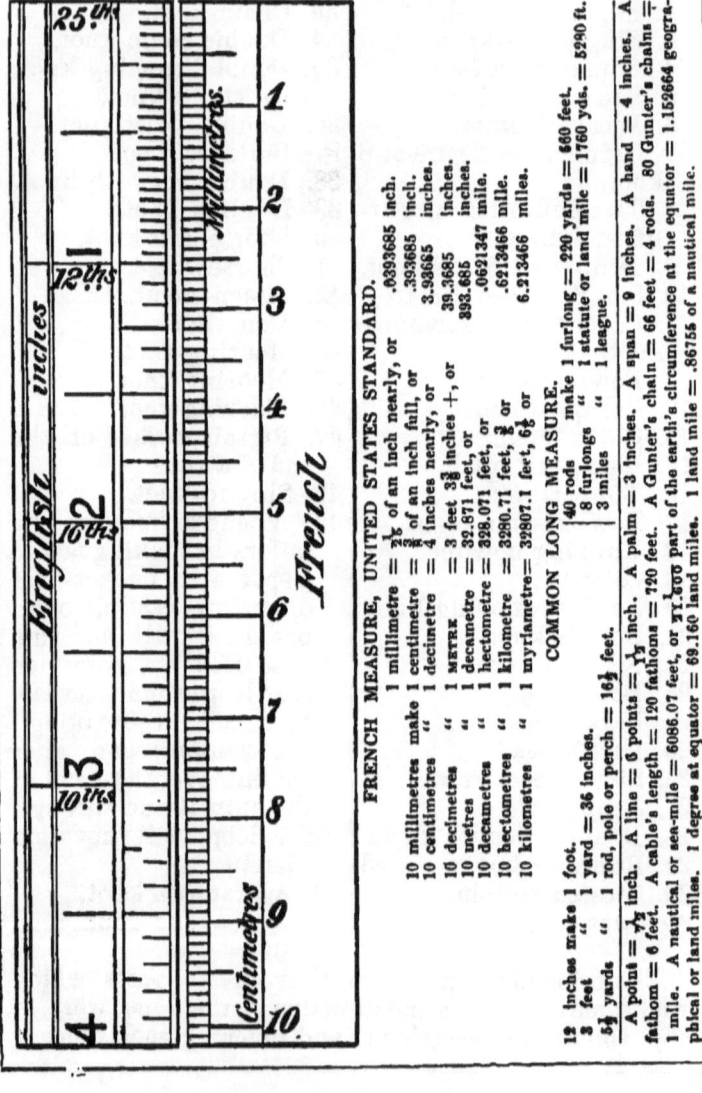

FRENCH MEASURE, UNITED STATES STANDARD.

10 millimetres make	1 millimetre = $\frac{1}{25}$ of an inch nearly, or	.0393685 inch.
10 centimetres "	1 centimetre = $\frac{2}{5}$ of an inch full, or	.393685 inch.
10 decimetres "	1 decimetre = 4 inches nearly, or	3.93685 inches.
10 metres "	1 METRE = 3 feet 3⅜ inches +, or	39.3685 inches.
10 decametres "	1 decametre = 32.871 feet, or	393.685 inches.
10 hectometres "	1 hectometre = 328.071 feet, or	.0621347 mile.
10 kilometres "	1 kilometre = 3280.71 feet, ⅝ or	.6213466 mile.
	1 myriametre = 32807.1 feet, 6⅛ or	6.213466 miles.

COMMON LONG MEASURE.

40 rods make	1 furlong = 220 yards = 660 feet.
8 furlongs "	1 statute or land mile = 1760 yds. = 5280 ft.
3 miles "	1 league.

12 inches make 1 foot.
3 feet " 1 yard = 36 inches.
5½ yards " 1 rod, pole or perch = 16½ feet.

A point = $\frac{1}{72}$ inch. A line = 6 points = $\frac{1}{12}$ inch. A palm = 3 inches. A span = 9 inches. A hand = 4 inches. A fathom = 6 feet. A cable's length = 120 fathoms = 720 feet. A Gunter's chain = 66 feet = 4 rods. 80 Gunter's chains = 1 mile. A nautical or sea-mile = 6086.07 feet, or $\frac{1}{21600}$ part of the earth's circumference at the equator = 1.159664 geographical or land miles. 1 degree at equator = 69.160 land miles. 1 land mile = .86755 of a nautical mile.

TABLES OF WEIGHTS

TROY WEIGHT.
24 grains............................1 pennyweight, dwt.
20 pennyweights................1 ounce = 480 grains.
12 ounces..........................1 pound = 5760 grains.
Troy weight is used for gold and silver.

APOTHECARIES' WEIGHT.
20 grains............................1 scruple.
3 scruples..........................1 dram = 60 grains.
8 drams.............................1 ounce = 480 grains.
12 ounces..........................1 pound = 5760 grains.

The ounce in the above measures, it will be noticed, is heavier, contains more grains than the common commercial or avoirdupois weight, but the avoirdupois pound is the heaviest.

AVOIRDUPOIS OR ORDINARY COMMERCIAL WEIGHT.
27.34 + grains....................1 dram.
16 drams...........................1 ounce = 437½ grains.
16 ounces..........................1 pound = 7000 grains.
28 pounds..........................1 quarter.
4 quarters..........................1 hundredweight = 112 lbs.
20 hundredweight..............1 ton = 2240 lbs.

The standard of the avoirdupois pound is the weight 27.7015 cubic inches of distilled water at 30°.85 F., barometer 30 inches.

A troy oz. = 1.09714 avoir. oz.
An avoir. oz. = .911458 troy oz.
A stone = 14 lbs. A quintal = 100 lbs.

FRENCH WEIGHTS (UNITED STATES STANDARD).

Grains avoir.
1 milligramme = .01543316.
10 milligrammes make 1 centigramme = .1543316.
10 centigrammes " 1 decigramme = 1.543316.
10 decigrammes " 1 GRAMME = 15.43316.

Pounds avoir.
10 grammes " 1 decagramme = .02204737.
10 decagrammes " 1 hectogramme = .2204737.
10 hectogrammes " 1 KILOGRAMME = 2.204737.
10 kilogrammes " 1 myriogramme = 22.04737.
10 myriogrammes " 1 quintal = 220.4737.
10 quintals " 1 tonne = 2204.737.

The gramme is the basis of the French weights, and consists of a cubic centimetre of distilled water.

SQUARE OR LAND MEASURE, UNITED STATES.
144 sq. inches = 1 sq. foot.
9 sq. feet = 1 sq. yard.
30¼ sq. yards = 1 sq. rod.
40 sq. rods = 1 sq. rood.
4 sq. roods = 1 sq. acre.

DRY MEASURE, UNITED STATES.
2 pints = 1 quart.
4 quarts = 1 gallon.
2 gallons = 1 peck.
4 pecks = 1 bushel.

TABLES OF MEASURES.

CUBIC OR SOLID MEASURE
1728 cubic ins. = 1 cubic foot.
27 cubic feet = 1 cubic yard.

LIQUID MEASURE, UNITED STATES.
4 gills = 1 pint.
2 pints = 1 quart.
4 quarts = 1 gallon.
63 gallons = 1 hogshead.

2 hogsheads = 1 pipe or butt.
2 pipes = 1 tun.

1 barrel = 31½ gallons.
1 gallon = 231 cubic inches.
1 bushel = 1.24445 cub. ft.
1 barrel flour = 196 lbs. = 3 bush.

A cylinder seven inches diameter and six inches high contains a gallon.

FRENCH SQUARE MEASURE, U. S.

Square inches.
1 square millimetre = .001549.
1 square centimetre = .154988.
1 square decimetre = 15.4988.

Square feet.
1 square metre = 10.763058.
1 square decametre = 1076.3058.
1 square decare = 10763.058.

U. S. acres.
1 square hectare = 2.47086.
1 square kilometre = 247.086.
1 square myriametre = 24708.6.

FRENCH CUBIC OR SOLID MEASURE, U. S.

Cubic inches.
Millitre or cubic centimetre .... = .0610165.
10 millitres make 1 centilitre = .610165.
10 centilitres " 1 decilitre = 6.10165.
10 decilitres " 1 litre = 61.0165.
10 litres " 1 decalitre = 610.165.

Cubic feet.
10 decalitres " 1 hectolitre = 3.53105.
10 hectolitres " 1 kilolitre or cubic metre = 35.3105.
10 kilolitres " 1 myriolitre = 353.105.

MEASURING LAND BY WEIGHT.

The area of any piece of land, no matter how irregular the boundary lines, may be accurately ascertained by means of a delicate balance as follows. Make a drawing of the plot of ground on pasteboard, to a given scale, say four square roods to one inch. Cut from some part of the sheet of pasteboard a piece exactly one inch square, which represents one acre or four square roods. Also cut out the plot as drawn. Weigh the square and the plot. The number of times the weight of the square is contained in the weight of the plot indicates the area of the land. For example, if the square which represents one acre weighs twenty grains, and the plot weighs two hundred and forty grains, then the plot contains twelve acres.

MOLECULES.

A MOLECULE is the smallest mass into which any substance can be subdivided without changing its chemical nature.

All substances are aggregations of isolated molecules.

A piece of gold having six plane surfaces, each one inch square, is called a cubic inch of gold, and looks as if it solidly filled that space. But it is not solid, for it is composed of individual molecules, which are separated by comparatively wide intervals.

Molecules are, to use the language of Sir William Thompson, "pieces of matter of measurable dimensions, with shape, motion, and laws of action." A molecule of glass, as measured by this philosopher, is one five hundred millionth part of an inch in diameter.

Equal volumes of all substances, when in a state of gas, and under like conditions, contain the same number of molecules.

The number of molecules in a cubic inch of any perfect gas, at 32° F. and 30 ins. barometer pressure, is one hundred thousand millions of millions of millions, or 10^{23}.

The molecules of bodies are never at rest, but have a constant motion. The molecules of a gas confined in a vessel have great energy, are always flying about with a high velocity, but in straight lines. They strike against each other and rebound, they drive against the inner walls of the vessel, and the force of this impact of the molecules of the gas against the walls we call the pressure of a gas.

At a barometer pressure of 30 inches, or 15 lbs. to the square inch, temperature 32° F., the molecules of hydrogen have a velocity of 6097 feet per second, or over 4000 miles per hour. The energy of a pound of hydrogen, under the above conditions, is equal to that of a cannon-ball of the same weight having the same velocity.

A cubic inch of water may by heat be expanded into gaseous form, or steam, occupying the space of a cubic foot. In both forms the same number of molecules of water are found; but in the gaseous condition, the molecules are much more widely separated than in the liquid; so widely, in fact, that a cubic foot of alcohol vapor together with a cubic foot of ether vapor may be introduced into the vessel—or, apparently, just as much of the alcohol, and just as much of the ether, as if there were no water vapor present. All these vapors remain separate; they do not chemically unite.

HOW TO INVENT.

IN order to succeed, a new invention must be superior to any thing that has preceded it, and must be sold at a price that will enable it to be brought into general use.

People can not afford to throw away old implements unless the new ones are sufficiently superior to make up for the loss. Let inventors produce a good article, at a moderate price, and they will be sure of success.

The readiest way to invent is to *keep thinking;* and to thought add *practical experiments.* Book knowledge is not essential. Examine things about you, note how they are made, and study how to improve them. Keep your eyes and ears open, ask questions, be a continual seeker after useful information. Those who do this, soon acquire a knowledge of the sciences, and insensibly become educated. Among the beneficent influences of the Patent laws is the fact that they incite the humblest individuals to study.

To avoid waste of time in reproducing old devices, the inventor should be well posted in regard to inventions that have already been made. He should also be informed as to the particular classes of devices in which improvements are most urgently demanded. For this purpose, an attentive study of THE SCIENTIFIC AMERICAN will be almost indispensable. This splendid newspaper is published weekly, and furnishes the latest information concerning the progress of new discovery, with elegant engravings. $3.00 a year. MUNN & CO., 361 Broadway, New York. Send to us for a specimen copy.

HISTORY OF THE STEAM-ENGINE.

PAPIN, of France, was the first (in 1690) to operate a piston by steam, which acted only on one side of the piston. He also invented the safety valve. He was born 1650, died 1710. Savery, 1697, first employed steam-power in doing useful work. His piston, like Papin's, took steam on one side only, the pressure of the atmosphere being admitted to the other side. James Watt was the first to make the complete steam-engine, or the existing forms in which steam acts on both sides of the piston. He also made the steam-condenser, the governor, the walking-beam, applied the fly-wheel, and nearly all the parts of the modern engine. He was born 1736, died 1819. He made a rotary steam engine in 1782, and patented a locomotive engine in 1784. In 1804, Trevithick and Vivian operated a locomotive which traveled five miles an hour, with a load of ten tons. Cook, in 1808, used fixed engines and ropes to draw railway-cars. Blachett and Hedley, in 1813, discovered that smooth locomotive wheels might be used on railways, instead of toothed wheels and toothed rails before required. George Stevenson, 1825, made railway locomotion successful by adapting the locomotive to variable speeds and loads, by means of his blast-pipe, and by introducing the tubular boiler, which latter was suggested to him and invented by Booth, 1829. October 6th, 1829, the famous competitive trial of locomotives on the Liverpool and Manchester railway took place, which established the superiority of Stevenson's locomotives. and inaugurated the art of railway communication.

The first steamboat actually employed in business was a small vessel built by John Fitch of Pennsylvania, 1790, worked on the Delaware ; speed, 7½ miles per hour. Robert Fulton's steamboat, the Clermont, made her first trip from New-York to Albany, August, 1807 ; speed, five miles per hour. The first steam-vessel to cross the Atlantic was the Savannah, in 1819, from Savannah to Liverpool, 26 days. In 1838 the Sirius arrived at New-York, 17 days from London ; and the Great Western, 15 days from Bristol.

HEAT.–ITS MECHANICAL EQUIVALENT.

HEAT is a peculiar motion of the particles of matter which prevents their contact. Heat and mechanical power are convertible forces. The force of the heat that raises one pound of water 1° F. will lift a weight of 772 lbs. one foot high. The power of a weight of 772 lbs. descending one foot, if applied to a small paddle wheel turning in one pound of water, will, by friction, raise the temperature of the water 1° F.

A heat-unit is the amount of heat that raises a pound of water 1° F., or that lifts a weight of 772 lbs. one foot high.

The mechanical equivalent of a heat-unit is the power of a weight of 772 lbs. descending one foot, or of a one-pound weight descending 772 feet. Hence,

$$772 \text{ foot-pounds} = 1 \text{ heat-unit},$$
$$1 \text{ heat-unit} = 772 \text{ foot-pounds}.$$

A galvanic battery that produces an electrial current capable of heating one pound of water 1° F., will yield magnetic force sufficient to raise a weight of 772 lbs. one foot high.

Thus heat, electricity, magnetism, and chemical force are brought into numerical correlation with mechanical power.

The illustrious philosopher, Dr. J. P. Joule, of Manchester, England, first promulgated the mechanical equivalent of heat, A.D. 1845.

COPYING-INK.

TAKE two gallons of rain-water, and put into it ¼ pound of gum arabic, ¼ pound brown sugar, ¼ pound clean copperas, ¾ pound powdered nut-galls. Mix and shake occasionally for ten days, and strain. If needed sooner, let it steep in an iron kettle until the strength is obtained.

VELOCITY AND FORCE OF THE WIND.

Miles per Hour.	Feet per Minute.	Pressure on a Square Foot in Pounds.	Description of the Wind.
1	88	.005	Barely observable.
2	176	.02	} Just perceptible.
3	264	.045	
4	352	.08	Light breeze.
5	440	.125	} Gentle, pleasant wind.
6	528	.18	
8	704	.32	
10	880	.5	Fresh breeze.
15	1320	1.125	Brisk blow.
20	1760	2.	Stiff breeze.
25	2200	3.125	Very brisk.
30	2640	4.5	} High wind.
35	3080	6.125	
40	3520	8.	Very high wind.
45	3960	10.125	Gale.
50	4400	12.5	Storm.
60	5280	18.	Great storm.
80	7040	32.	Hurricane.
100	8800	50.	Tornado.

GUNPOWDER.

The heat developed at the moment of explosion is 4664° Fahr., and the resulting gas pressure, if the powder closely fills the chamber, is 40 tons or 80,000 lbs. to the square inch.

Careful experiment by De Saint Robert with rifled cannon of 3¼ inches bore, 8½ lbs. shell, 1¼ lbs. powder, gives 1300 ft. velocity per second, or a little over 900 miles per hour, for the shell when it leaves the mouth of the cannon, which is equal to a force of 219,000 foot-pounds, or a little less than seven horse-power. But the heat actually developed by the above amount of powder corresponds to almost thirty-two horse-power of work; seventy-nine per cent of the power of the powder is therefore lost.

SPECIFIC GRAVITY.

SPECIFIC GRAVITY AND WEIGHT OF VARIOUS SUBSTANCES. WATER = 1.

	Specific gravity.	Weight per cubic foot, lbs.	Weight per cubic inch, lbs.
Acetic acid............	1.06	66.	.038
Alcohol	.792	49.	.028
Aluminium, sheet......	2.67	166.6	.096
Antimony, cast........	6.72	419.5	.242
Asb, dry..............	.69	43.	.025
" green............	.76	47.	.027
Asphalt...............	2.5	156.	.09
Basalt................	2.95	184.	.106
Beech, dry............	.69	43.	.025
Bell-metal............	8.05	502.52	.29
Birch.................	.69	43.	.025
Bismuth, cast.........	9.822	613.1	.353
Box...................	1.28	80.	.046
Brass, cast...........	8.4	524.37	.3
" sheet............	8.44	526.86	.301
Brick, common.. { from	1.6	100.	.09
{ to	2.	125.	.057
Cedar, American.......	.554	35.	.020
" Lebanon........	.486	30.	.017
" West-Indian.....	.748	48.	.026
" Indian..........	1.315	82.15	.
Cement, Portland......	1.4	87.	.05
" Roman..........	1.6	100.	.057
Chalk.................	2.33	145.	.084
Chestnut..............	.606	38.	.022
Clay..................	1.9	119.	.068
Coal, anthracite......	1.53	95.	.055
" bituminous.......	1.27	79.	.045
Coke..................	.744	46.	.026
Concrete, ordinary....	1.9	119.	.068
" in cement.....	2.2	137.	.079
Cork..................	.240	15.	.008
Copper, cast..........	8.607	537.3	.31
" sheet...........	8.78	548.1	.316
Deal, Norway..........	.689	43.	.025
Earth........... { from	1.52	77.	.054
{ to	2.00	125.	.072
Ebony.................	1.187	74.	.043
Elm...................	.579	36.	.021
" Canadian..........	.725	45.	.026
Ether.................	.716	45.	.026
Fir, spruce...........	.512	32.	.018
Firestone.............	1.8	112.	.065
Glass, flint..........	3.078	192.	.111
" crown...........	2.52	157.	.091
" common green...	2.52	159.	.091
" plate............	2.76	172.	.099
Gold..................	19.36	1208.5	.697
Granite...............	2.65	165.75	.096
Gun metal [10 cop.,1 tin]	8.561	534.42	.308
Gutta-percha..........	.966	60.	.035
Gypsum................	2.286	143.	.082
Hornbeam..............	.76	47.	.027
Hydrochloric acid.....	1.2	75.	.043
Iron, cast, average...	7.23	451.	.26
" wrought, average..	7.78	485.6	.28
Ironwood..............	1.15	71.	.041
India-rubber..........	.93	58.	.033

THE specific gravity of any liquid or solid body is its weight as compared with an equal volume of pure water at 60° F. Water = 1.

THE specific gravity of a gas is its weight as compared with an equal volume of pure air at 60°. F. Air = 1.

SPECIFIC GRAVITY OF GASES.

$Air = 1$.

Hydrogen...........0.0692
Steam 0.4883
Marsh gas..........0.5596
Carbonic oxide......0.967
Nitrogen...........0.9713
Oxygen.............1.1057
Carbonic acid..... 1.529
Sulphurous acid....2.25
Chlorine 2.47

SPECIFIC HEAT.

IF 1 lb. of water, 1 lb. of mercury, 1 lb. of silver, 1 lb. of iron be exposed to a heat sufficient to raise the water 1° F., the temperature of the mercury will be found to be 30°, of the silver 17.5°, the iron 8.8°. The specific heat of different substances is found by comparing their temperature with water as above. Thus, the specific heat of water is 1; the specific heat of mercury is $\frac{1}{30}$, or one-thirtieth that of water; silver, $\frac{1}{17.5}$, iron, $\frac{1}{8.8}$.

FRICTION.

A BAG of wheat weighing 200 lbs. is dragged on the floor by means of a spring-balance, the pointer of which indicates 40 lbs. as the force required to move the bag. Make that force, 40, the numerator of a fraction, and the moved weight, 200, the denominator. Then $\frac{40}{200}$ or $\frac{1}{5}$ of the weight is the *co-efficient of friction*, or the force required to overcome the friction. $\frac{1}{5}$ of 200 lbs. is 40 lbs., which is the force indicated in this example to overcome the friction of the bag. If the load or weight were 400 lbs., and the co-efficient of friction $\frac{1}{4}$, then it would take $\frac{1}{4}$ of 400 lbs., or 100 lbs. force, to move the load.

SPECIFIC GRAVITY. 131

Specific Gravity and Weight of Various Substances. Water = 1.

	Specific gravity.	Weight per cubic foot, lbs.	Weight per cubic inch, lbs.
Ivory	1.82	114.	.065
Larch	.543	34.	.019
Lead, cast	11.36	708.5	.408
" sheet	11.4	711.6	.41
Lignum vitæ	1.333		.048
Lime-wood	.564	35.	.02
Lime, quick	.843	53.	.03
Limestone	3.180	198.75	
Logwood	.913	57.06	
Mahogany, Honduras	.560	35.	.02
" Nassau	.668	42.	.024
" Spanish	.852	53.	.031
Maple	.675	42.	.025
Marble	2.72	170.	.098
Mercury	13.596	848.75	.489
Mortar, average	1.7	106.	.061
Muriatic acid	1.2	75.	
Nitric acid	1.217	75.	.044
Oak, African	.988	62.	.035
" American, red	.85	53.	.03
" " white, dry	.779	49.	.028
" Canadian	.872	54.5	
" English, white, dry	.777	48.	.028
" " " green	.934	58.	.034
" live, seasoned	1.068	66.75	
" " green	1.260	78.75	
Oil, linseed	.94	58.	.034
" olive	.915	57.	.033
" whale	.923	58.	.033
Oolite, Portland stone	2.423	151.	.087
" Bath stone	1.978	123.	.072
Pine, red, dry	.590	37.	.022
" white, dry	.554	34.62	
" yellow, dry	.461	28.81	
" pitch	.660	41.25	
Pitch	1.15	69.	.041
Platinum, average	21.531	1343.9	.775
Plumbago	2.267	140.	.082
Salt	2.13	133.	
Sand, quartz	2.75	171.	.099
" river	1.88	117.	.067
" fine	1.52	95.	.054
" coarse	1.61	100.	.058
Sandstone			
Satinwood	.96	60.	.034
Silver	10.474	653.8	.377
Slate	2.88	180.	.104
Sulphur			
Sulphuric acid	1.84	115.	.066
Tallow	.94	59.	.034
Tar	1.01	63.	.036
Teakwood	.806	50.	.028
Tile, average	1.83	113.50	.065
Tin, cast	7.29	45.51	.262
Water, distilled, 39°	1.000	62.425	.036
" sea	1.027	64.	.037
White metal (Babbitt)	7.31	456.32	.263
Zinc, cast	7.	437.	.252

Iron bars, one inch square, rolled, weigh 3.38 lbs. per foot; round bars, one inch diameter, 2.65 lbs. per foot.

From 300 to 324 cubic feet of dry clover, or 216 to 243 cubic feet of dry hay, weigh a ton. 270 cubic feet of new hay in a ton.

In this country the average weight of men is 141 1-2 lbs.; women, 124 1-2 lbs.

The weight of horses in this country is from 800 to 1200 lbs.

The standard weight of a bushel of wheat is 60 lbs.; corn and rye, 56 lbs.; oats, 32 lbs.; barley, 48 lbs.

Potatoes, in weight 100 lbs., are made up of 75.9 lbs. water; albumen, 2.3 lbs.; oily matter, 0.2 lbs.; woody fibre, 0.4 lbs.; starch, 20.2 lbs.; minerals, 1 lb.

A kind of tracing paper, which is transparent only temporarily, is made by dissolving castor-oil in absolute alcohol and applying the liquid to the paper with a sponge. The alcohol speedily evaporates, leaving the paper dry. After the tracing is made, the paper is immersed in absolute alcohol, which removes the oil, restoring the sheet to its original opacity.

The diameter of a barrel at the heads is 17 inches; bung, 19 inches; length, 26 inches; volume, 7689 cubic inches.

INCUBATION.—The temperature of hatching eggs is 104° F. Periods: swan, 42 days; parrot, 40; goose and pheasant, 35; duck, turkey, peafowl, 28; hens, 21; pigeons, 14; canary birds, 14.

PERIODS OF GESTATION.—Guinea-pig, 3 weeks; sow, 16 weeks; cat, 8 weeks; dog, 9 weeks; lion, 5 months; sheep, 5 months; cow, 9 months; horse and ass, 11 months; buffalo and camel, 12 months; elephant, 23 months.

SMALL STEAMBOATS.

THE following is an example of the practical way in which special questions put by subscribers to the *Scientific American* are answered by the editors:

"H. C. E. says: 1. I have a boat, 21 feet long by 7 feet 6 inches beam, drawing 12 or 15 inches of water. I built an engine 3x5 inches, with a link motion. Is the engine large enough for the boat? A. Yes. 2. I have a ½-inch feed-pipe and ¾-inch exhaust. Is the exhaust too small for the engine? A. It will answer very well. 3. What size of propeller should I use? A. Of 18 or 20 inches diameter, 2½ feet pitch. 4. What size of boiler is required? A. About 2½ feet diameter, 4 feet high. 5. What is meant by the pitch of a propeller? A. It is the distance it would move the boat, at each revolution, if it worked in an unyielding medium, like a screw in a nut."

The foregoing is an epitome of dimensions sufficient to enable any intelligent machinist to build a fast and serviceable steamer. Hundreds of these little vessels are now in use throughout the country, upon the smaller lakes and shallow streams. Their use has become greatly extended by the publication of practical details of construction and management in the *Scientific American*.

Queries relating to steam engineering are answered in the *Scientific American* by an experienced engineer; those pertaining to electricity, by a practical electrician; chemical inquiries, by a superior chemist; mechanical questions, by a talented machinist; astronomical inquiries, by an astronomer; and so on, for nearly all of the departments of science. The amount of valuable information thus made public through the columns of the *Scientific American* is very large. It is, on this account, the most positively valuable weekly newspaper ever published.

COPPER, if suddenly cooled, becomes soft and malleable; if slowly cooled it hardens and becomes brittle.

PRACTICAL GEOMETRY.

A KNOWLEDGE of geometry, both practical and theoretical, is of importance to mechanics and inventors. It is promotive of truth and patience in mental habits, and leads to the exercise of nicety and exactness in the execution of mechanical labors. With a pair of dividers, a rule and pencil, any person may speedily acquire a considerable knowledge of practical geometry. We subjoin a few simple and generally useful problems for practice, in the hope of thus interesting some of our readers in the subject, so that they will continue the study. Complete works on geometry can be had at the book-stores.

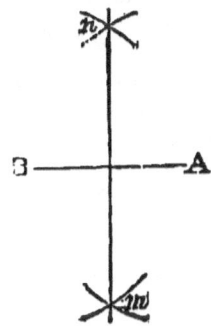

Problem 1.—To divide a line into equal parts.—To draw a line perpendicular to another: With a pair of dividers from the extremities of the line A B as centres, with any distance exceeding the point where the line is to be intersected, describe arcs cutting each other as *m n;* then a line drawn through *m n* will divide the line A B equally, and will also be perpendicular thereto.

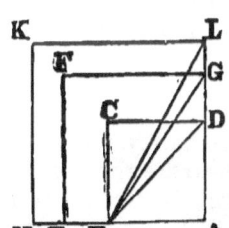

Problem 2.—To find the side of a square that shall be any number of times the area of a given square: Let A B C D be the given square; then will the diagonal B D be the side of a square A E F G, double in area to the given square A B C D; the diagonal B D is equal to the line A G; if the diagonal be drawn from B to G, it will be the side of a square A H K L, three times the area of the square A B C D; the diagonal B L will equal the size of a square four times the area of the square A B C D, etc.

PRACTICAL GEOMETRY.

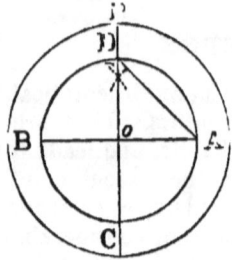

Problem 3.—To find the diameter of a circle that shall be any number of times the area of a given circle: Let A B C D be the given circle; draw the two diameters A B and C D at right angles to each other, and the cord A D will be the radius of the circle *o* P, twice the area of the given circle nearly; and half the cord will be the radius of a circle that will contain half the area, etc.

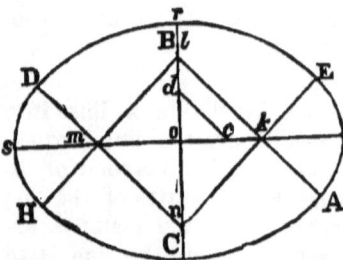

Problem 4.—To describe an ellipsis, the transverse and conjugate diameters being given: From *o*, as a centre, with the difference of the transverse and conjugate semi-diameters, set off *o c* and *o d ;* draw the diagonal *c d*, and continue the line *o c* to *k*, by the addition of half the diagonal *c d*, then will the distance *o k* be the radius of the centres that will describe the ellipsis; draw the lines A B, C D, C E, and B H, cutting the semi-diameters of the ellipsis in the centres *k* B *m n ;* then with the radius *m s*, and with *k*, and *m* as centres, describe the arcs D H and A E; also, with the radius *n r*, and with *n* and B as centres, describe the arcs E H and A H, and the figure A E D H will be the ellipsis required.

THE "SCIENTIFIC AMERICAN."—"It is hardly necessary for us to speak of its merits to those who are thoroughly posted up in the improvements of the age; but the general reading public may not be so well aware that it contains the finest engravings of all the late inventions—the new monitors, army and navy weapons, vessels, forts, machinery of all kinds, military and civil, mechanical and agricultural—with essays from the most distinguished scholars.

THE STEAM-ENGINE.

EVERY mechanic and inventor should make himself generally familiar with the construction and operation of the steam-engine. To assist them in gaining this knowledge, we subjoin for reference a diagram of the common Condensing Engine, with letters of reference to the names of the various parts:

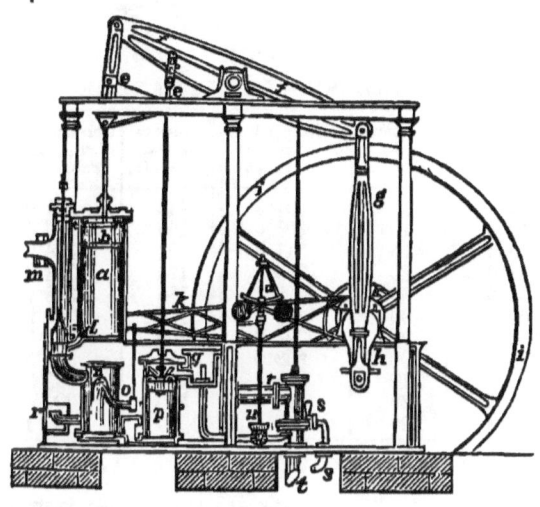

a, steam cylinder; b, piston; c, upper steam port or passage; d, lower steam port; $e\,e$, parallel motion; $f\,f$, beam; g, connecting rod; h, crank; $i\,i$, fly-wheel; $k\,k$, eccentric and its rod for working the steam-valve; l, steam-valve and casing; m, throttle-valve; n, condenser; o, injection-cock; p, air-pump; q, hot well; r, shifting-valve to create vacuum in condenser previous to starting the engine; s, feed-pump to supply boilers; t, cold-water pump to supply condenser; u, governor. A study of the above diagram and description, in connection with attentive observation of engines in motion, will be of much assistance in acquiring a general understanding of the machine. We recommend the following standard works for careful study by all who desire to become thoroughly posted: Bourne's Catechism of the Steam Engine, Main & Brown's Marine Steam Engine.

EFFECTS OF HEAT UPON BODIES.

	Deg. Fah.		Deg. Fah.
Cast iron melts	2,786	Cadmium	450
Gold "	2,016	Tin melts	442
Copper "	1,996	Tin and bismuth, equal	
Brass "	1,900	parts, melt	283
Silver "	1,873	Tin 3 parts, bismuth 5	
Red heat visible by day	1,077	parts, lead 2 parts, melt	212
Iron red hot in twilight	884	Sodium	190
Common fire	790	Alcohol boils	174
Zinc melts	773	Potassium	136
Iron, bright red in dark	752	Ether boils	98
Mercury boils	630	Human blood (heat of)	98
Lead melts	612	Strong wines freeze	20
Linseed oil boils	600	Brandy freezes	7
Bismuth melts	497	Mercury freezes	−39½

STEAM PRESSURE AND TEMPERATURE.

Pressure in lbs. per sq. in.	Corresponding Temperature Fahrenheit.	Pressure in lbs. per sq. in.	Corresponding Temperature. Fahrenheit.	Pressure in lbs. per sq. in.	Corresponding Temperature. Fahrenheit.
10	192·4	65	301·3	140	357·9
15	212·8	70	306·4	150	363·4
20	228·5	75	311·2	160	368·7
25	241·0	80	315·8	170	373·6
30	251·6	85	320·1	180	378·4
35	260·9	90	324·3	190	382·9
40	269·1	95	328·2	200	387·3
45	276·4	100	632·0	210	391·5
50	283·2	110	339·2	220	395·5
55	289·3	120	345·8	230	399·4
60	295·6	130	352·1	240	403·1

HEAT AND ELECTRICAL CONDUCTIVITY.

Substances.	Heat Conductivity.	Electrical Conductivity.
Silver	100·0	100·0
Copper	73·6	73·3
Gold	53·2	58·5
Brass	23·6	21·5
Zinc	19·9	....
Tin	14·5	22·6
Steel	12·0	....
Iron	11·9	13·0
Lead	8·5	10·7
Platinum	6·4	10·3
Palladium	6·3	....
Bismuth	1·8	1·9

TABLE OF OCCUPATIONS,
COMPILED FROM THE CENSUS OF THE UNITED STATES, A. D. 1880.

All occupations (persons engaged in)...............	17,392,099
AGRICULTURE........total,	7,670,493
Agricultural laborers...........	3,323,876
Apiarists....................	1,016
Dairymen and dairywomen....	8,948
Farm and plantation overseers.	3,106
Farmers and planters.........	4,225,945
Florists.....................	4,550
Gardeners, nurserymen, and vine-growers..............	51,482
Stock-drovers................	3,449
Stock-herders................	24,098
Stock-raisers................	16,528
Turpentine farmers and laborers..................	7,450
Others in agriculture.........	45
PROFESSIONAL AND PERSONAL SERVICES.................	4,074,238
Actors.......................	4,812
Architects	3,375
Artists and teachers of art.....	9,104
Auctioneers..................	2,331
Authors, lecturers, and literary persons..................	1,131
Barbers and hairdressers.......	44,851
Billiard and bowling-saloon keepers and employes.......	1,543
Boarding and lodging-house keepers....................	19,058
Chemists, assayers, and metallurgists....................	969
Clergymen	64,698
Clerks and copyists (not otherwise described)...............	25,467
Clerks in government offices....	16,849
Clerks in hotels and restaurants	10,916
Collectors and claim agents....	4,213
Dentists.....................	12,314
Designers, draughtsmen, and inventors...................	2,820
Domestic servants..............	1,075,655
Employés of charitable institutions........................	2,396
Employés of government (not clerks).....................	31,601
Employés of hotels and restaurants (not clerks).............	77,413
Engineers (civil)................	8,261
Hostlers......................	31,697

PROFESSIONAL AND PERSONAL.	
Hotel keepers.................	32,453
Hunters, trappers, guides, and scouts.....................	1,912
Janitors......................	6,763
Journalists...................	12,308
Laborers (not specified)........	1,859,223
Launderers and laundresses...	121,942
Lawyers......................	64,137
Livery stable keepers..........	14,213
Messengers...................	13,985
Midwives	2,118
Musicians (professional) and teachers of music............	30,477
Nurses.......................	13,483
Officers of the army and navy (United States)...............	2,600
Officials of government........	67,081
Physicians and surgeons........	85,671
Restaurant keepers............	13,074
Sextons......................	2,449
Showmen and employés of shows......................	2,604
Soldiers, sailors, and marines (United States army and navy)	24,161
Teachers and scientific persons..	227,710
Veterinary surgeons...........	2,130
Watchmen (private) and detectives....................	13,394
Whitewashers.................	3,316
Others in professional and personal services..............	4,570
TRADE AND TRANSPORTATION..	1,810,256
Agents (not specified).........	18,523
Bankers and brokers of money and stocks.................	15,180
Boatmen and watermen.......	20,368
Bookkeepers and accountants in stores.....................	59,790
Brokers (commercial)..........	4,193
Canalmen....................	4,329
Clerks in stores...............	353,444
Clerks and bookkeepers in banks	10,267
Clerks and bookkeepers in express companies.............	1,856
Clerks and bookkeepers in insurance offices...............	2,830
Clerks and bookkeepers in railroad offices...................	12,321
Commercial travellers..........	28,158
Draymen, hackmen, teamsters, etc........................	177,586
Employés in warehouses........	5,022

TABLE OF OCCUPATIONS.

TABLE OF OCCUPATIONS. (*Continued.*)

TRADE AND TRANSPORTATION.

Occupation	Number
Employés of banks (not clerks)	1,070
Employés of insurance companies (not clerks)	13,146
Employés of railroad companies (not clerks)	236,058
Hucksters and peddlers	53,491
Milkmen and milkwomen	9,242
Newspaper criers and carriers	3,374
Officials and employés of express companies (not clerks)	13,004
Officials and employés of street railroad companies	11,925
Officials and employés of telegraph companies	22,809
Officials and employés of telephone companies	1,197
Officials and employés of trading and transportation companies (not specified)	9,702
Officials of banks	4,421
Officials of insurance companies	1,774
Officials of railroad companies	2,069
Packers	4,176
Pilots	3,770
Porters and laborers in stores and warehouses	32,192
Sailors	60,070
Salesmen and saleswomen	32,279
Saloon keepers and bartenders	68,461
Shippers and freighters	5,166
Steamboat men and women	12,365
Stewards and stewardesses	2,283
Toll-gate and bridge keepers	2,303
Traders and dealers (not specified)	112,840
Traders and dealers in agricultural implements	1,999
Traders and dealers in books and stationery	4,982
Traders and dealers in boots and shoes	9,993
Traders and dealers in cabinet ware	7,419
Traders and dealers in cigars and tobacco	11,866
Traders and dealers in clothing and men's furnishing goods	10,073
Traders and dealers in coal and wood	10,871
Traders and dealers in cotton and wool	2,494
Traders and dealers in crockery, china, glass, and stone ware	2,373
Traders and dealers in drugs and medicines	27,700
Traders and dealers in dry goods, fancy goods, and notions	45,831
Traders and dealers in gold and silver ware and jewelry	2,305

TRADE AND TRANSPORTATION.

Occupation	Number
Traders and dealers in groceries	101,849
Traders and dealers in hats, caps, and furs	4,809
Traders and dealers in ice	2,854
Traders and dealers in iron, tin, and copper ware	15,076
Traders and dealers in junk	3,574
Traders and dealers in leather, hides, and skins	2,382
Traders and dealers in liquors and wines	13,500
Traders and dealers in live-stock	12,596
Traders and dealers in lumber	11,363
Traders and dealers in marble, stone, and slate	1,405
Traders and dealers in music and musical instruments	1,906
Traders and dealers in newspapers and periodicals	2,729
Traders and dealers in oils, paints, and turpentine	1,940
Traders and dealers in paper and paper stock	1,862
Traders and dealers in produce and provisions	35,129
Traders and dealers in real estate	11,253
Traders and dealers in sewing machines	6,577
Undertakers	5,113
Weighers, gaugers, and measurers	3,302
Others in trade and transportation	177

MANUFACTURING, MECHANICAL, AND MINING INDUSTRIES	3,837,112
Agricultural implement makers	4,891
Artificial flower makers	3,399
Apprentices to trades	44,170
Bag makers	1,408
Bakers	41,309
Basket makers	5,654
Blacksmiths	172,726
Bleachers, dyers, and scourers	8,222
Blind, door, and sash makers	4,946
Boat makers	2,063
Bone and ivory workers	1,898
Bookbinders and finishers	13,833
Boot and shoe makers	194,079
Bottlers and mineral-water makers	2,081
Box factory operatives	15,762
Brass founders and workers	11,568
Brewers and maltsters	16,278
Brick and tile makers	36,052

TABLE OF OCCUPATIONS.

TABLE OF OCCUPATIONS. (Continued.)

Manufacturing, Mechanical, and Mining.		Manufacturing, Mechanical, and Mining.	
Bridge builders and contractors.	2,587	Lead and zinc works operatives	2,105
Britannia and japanned ware makers.	1,375	Leather case and pocket-book makers.	1,397
Broom and brush makers.	8,479	Leather curriers, dressers, finishers, and tanners.	29,842
Builders and contractors (not specified).	10,804	Lumbermen and raftsmen.	30,651
Butchers.	76,241	Machinists.	101,130
Button-factory operatives.	4,872	Manufacturers.	44,019
Cabinet makers.	50,654	Marble and stone cutters.	32,842
Candle, soap, and tallow makers.	2,923	Masons (brick and stone).	102,473
Car makers.	4,708	Meat and fruit preserving establishment employés.	2,860
Carpenters and joiners.	373,143	Meat packers, curers, and picklers.	3,436
Carpet makers.	17,068	Mechanics (not specified).	7,858
Carriage and wagon makers.	49,881	Mill and factory operatives (not specified).	30,836
Charcoal and lime burners.	5,851	Millers.	53,440
Cheese makers.	4,570	Milliners, dressmakers, and seamstresses.	285,401
Chemical-works employés.	2,923	Miners.	234,228
Cigar makers.	56,599	Mirror and picture frame makers.	2,503
Clerks and bookkeepers in manufacturing establishments.	10,114	Nail makers.	5,803
Clock and watch makers and repairers.	13,820	Officials of manufacturing and mining companies.	8,198
Confectioners.	13,692	Oil mill and refinery operatives	3,929
Coopers.	49,138	Oil well operatives and laborers	7,340
Copper workers.	2,342	Organ makers.	2,437
Corset makers.	4,660	Painters and varnishers.	128,556
Cotton-mill operatives.	169,771	Paperhangers.	5,013
Distillers and rectifiers.	3,245	Paper mill operatives.	21,430
Employés in manufacturing establishments (not specified).	34,536	Pattern makers.	5,822
Engineers and firemen.	79,628	Photographers.	9,990
Engravers.	4,577	Piano forte makers and tuners.	5,413
Fertilizer establishment operatives.	1,383	Plasterers.	22,083
File makers, cutters, and grinders.	1,839	Plumbers and gasfitters.	19,383
Fishermen and oystermen.	41,352	Potters.	7,233
Flax dressers.	1,894	Printers, lithographers, and stereotypers.	72,726
Fur workers.	1,580	Print works operatives.	5,419
Galloon, gimp, and tassel makers.	2,235	Publishers of books, maps, and newspapers.	2,781
Gas-works employés.	4,695	Pump makers.	1,366
Gilders.	1,763	Quarrymen.	15,169
Glass-works operatives.	17,934	Qdartz and stamp mill operatives and laborers.	1,449
Glove makers.	4,511	Rag pickers.	2,206
Gold and silver workers and jewelers.	28,405	Railroad builders and contractors.	1,206
Gun and locksmiths.	10,572	Roofers and slaters.	4,026
Hair cleaners, dressers, and workers.	1,965	Rope and cordage makers.	3,514
Harness and saddle makers.	39,960	Rubber factory operatives.	6,350
Hat and cap makers.	16,860	Sail and awning makers.	2,950
Hosiery and knitti g mill operatives.	12,194	Salt makers.	1,431
Iron and steel works and shops operatives.	114,539	Saw and planing mill operatives.	77,050
Lace makers.	1,708	Sawyers.	5,195

TABLE OF OCCUPATIONS.

TABLE OF OCCUPATIONS. (*Continued.*)

MANUFACTURING, MECHANICAL, AND MINING.		MANUFACTURING, MECHANICAL, AND MINING.	
Scale and rule makers	1,027	Thread makers	3,259
Screw makers	1,361	Tinners and tinware makers	42,818
Sewing machine factory operatives	2,725	Tool and cutlery makers	13,749
		Trunk, valise, and carpet bag makers	3,013
Sewing machine operators	7,505		
Shingle and lath makers	5,166	Tobacco factory operatives	20,446
Ship carpenters, calkers, riggers, and smiths	17,452	Umbrella and parasol makers	1,967
		Upholsterers	10,443
Shirt, cuff, and collar makers	11,823	Wheelwrights	15,592
Silk mill operatives	18,071	Wire makers and workers	7,170
Starch makers	1,385	Wood choppers	12,731
Stave, shook, and heading makers	4,061	Wood turners, carvers, and woodenware makers	12,964
Steam boiler makers	12,771	Woolen mill operatives	86,010
Stove, furnace, and grate makers	3,341	Others in manufacturing, mechanical, and mining industries	13,542
Straw workers	4,229		
Sugar makers and refiners	2,327		
Tailors and tailoresses	133,756		

GENERAL SUMMARY.

Engaged in agriculture... 7,670,493
" " professional and personal services........................... 4,074,238
" " trade and transportation... 1,810,256
" " manufacturing, mechanical, and mining industries..... 3,837,112

All occupations 1880, persons engaged in.................................17,392,099

AERIAL NAVIGATION.

No motive power machine sufficiently light and powerful to lift itself from the ground and maintain itself in the air for any considerable time has yet been invented. Aerial navigation is therefore at present limited to the use of balloons, filled with light gas or hot air. Common coal gas is found to be the cheapest and most generally available gas for ballooning. 1000 cubic feet of coal gas will lift 35 lbs. weight. But hydrogen is the best gas for the purpose. 1000 cubic feet of hydrogen gas will lift from 60 to 70 lbs. It is the lightest of all substances. It is fifteen times lighter than air, and over eleven thousand times lighter than water. One of the cheapest ways to make hydrogen for balloons is to dissolve zinc in sulphuric acid; the latter is composed of sulphur and hydrogen. When the acid is poured on zinc, the sulphur unites with the metal and sets free the hydrogen, which bubbles up, and is conducted in a pipe to the balloon. Various efforts to propel and steer balloons have been made, by means of propellers turned by hand; also by the use of the electrical storage battery. Balloons are generally made of cotton cloth or silk, varnished with linseed oil, and dissolved rubber is sometimes mixed with the oil.

LIGHTNING RODS.

THE golden rule of safety is to provide the building with plenty of rods or conductors, and make sure that their lower ends, in the ground, are soldered to a *large surface* of metal or other conducting material placed underground. The common method is simply to stick the end of the rod three feet down into dry earth. But this is unsafe, because dry earth is a very poor conductor of electricity. To compensate for this lack of conductivity, the bottom of the rod should communicate with a large extent of conducting material. This may be done by soldering the lower end of the rod to water or gas pipes, if they exist; if not, then dig a trench four feet deep, six inches wide at bottom, in which deposit a layer six inches thick of coal dust, charcoal, anthracite, or bituminous. Bed the end of the rod for several feet in this layer of coal, which is a pretty good conductor of electricity. The trench should be one hundred or more feet in length. The rods should be placed on all ridges, corners, and chimneys. All metallic roofs, gutters, and water-pipes should have soldered connection with the rod, which should not be insulated from the building. All metallic water, gas, stove, and other pipes within the building, all bell wires, masses of metal, machinery, should be connected with the ground, in the manner described for the rods. The more conductors that are connected with the ground in this way, the greater the safety. Common one half inch square iron makes a good conductor. So does copper wire one eighth inch in diameter.

UNDERDRAINING.—Surface water that flows off the land instead of passing through the soil, carries with it whatever fertilizing matter it may contain, and abstracts some from the earth. If it pass down through the soil to drain, this waste is arrested.

COMMON hydraulic cement mixed with oil forms a good paint for roofs and out-buildings. It is water-proof and incombustible.

ORDINARY CONCRETE is a coarse mortar made of 1 measure of fine quicklime, 1½ of water, 6 to 8 of gravel and broken bricks or stones. Forms a cheap foundation or wall. The ground trench being made, the concrete, well mixed, is thrown in, in layers of one foot thickness, and thoroughly compacted by rammers. Above ground, wide planks are used to form trench. If carefully mixed and rammed, durable walls may be quickly and cheaply made.

MINERAL CONSTITUENTS ABSORBED OR REMOVED FROM AN ACRE OF SOIL BY THE FOLLOWING CROPS.

	Wheat, 25 bushels.	Barley, 40 bushels.	Turnips, 20 tons.	Hay, 1½ tons.
	Lbs.	Lbs.	Lbs.	Lbs.
Potassa	29.6	17.5	47.1	38.2
Soda	3.	5.2	8.2	12.
Lime	12.9	17.	29.9	44.5
Magnesia	10.6	9.2	19.7	7.1
Oxide of Iron	2.6	2.1	7.1	.6
Phosphoric Acid	20.6	25.8	46.3	15.1
Sulphuric Acid	10.6	2.7	13.3	9.2
Chlorine	2.	16.	3.6	4.1
Silica	118.1	129.5	247.8	78.2
Alumina	....	2.4	...	....
Total	210.00	213.00	423.00	209.00

USEFUL FACTS AND RECIPES.

FREEZING MIXTURES.—A mixture of nine parts phosphate of soda, six parts nitrate of ammonia, and four parts nitric acid, is a freezing compound which will cause a fall in temperature of 71° Fahrenheit.

Equal weights of sal-ammoniac and nitre, dissolved in its own weight of water, lowers the temperature of the latter from 50° Fahrenheit to 10° or 22° below the freezing point.

SOAP-BUBBLES.—Few things amuse children more than blowing bubbles. Dissolve a quarter of an ounce of Castile or oil soap, cut up in small pieces, in three quarters of a pint of water, and boil it for two or three minutes; then add five ounces of glycerine. When cold, this fluid will produce the best and most lasting bubbles that can be blown.

LIQUID GLUE.—Soak isinglass or refined gelatine in cold water until quite soft, and then dissolve it by trituration in the smallest possible quantity of proof spirit (49 per cent alcohol) over a water-bath. In 2 ounces of this dissolve by trituration, as before, 10 grains of gum ammonicum, and, while still liquid, add ½ drachm of mastic dissolved in 3 drachms of rectified spirit (84 per cent alcohol). Continue the trituration until the mixture is perfectly clear, then bottle. The cement is usually quite liquid in summer; for use in cold weather stand the bottle in warm water for a few minutes. This cement is quite colorless when properly made and improves by age. Joints made with this preparation are exceedingly strong and not affected by moisture.

CIRCLES.

A CIRCLE is the most capacious of all plain figures, or contains the greatest area within the same outline or perimeter.

To find the circumference of a circle, multiply the diameter by 3.1416, and the product will be the circumference.

To find the diameter of a circle, divide the circumference by 3.1416, and the quotient will be the diameter.

Any circle whose diameter is double that of another, contains four times the area oт the other.

To find the area of an ellipsis, multiply the long diameter by the short diameter and by .7854; the product will be the area.

To find the area of a circle, multiply the square of the diameter by the decimal .7854. Or multiply the circumference by the radius, and divide the product by 2.

PLASTER OF PARIS may be prepared in three ways to form a hard casting capable of taking a high polish. Mixed with a solution of borax, then rebaked, powdered, and mixed with a solution of alum, it forms Parian. Mixed with solution of alum, rebaked, powdered, and mixed with alum solution, it forms Keene's cement. Mixed with solution of sulphate of potash, rebaked, powdered, and mixed with alum solution, it forms Martin's cement.

WATER is the offspring of the two gases hydrogen and oxygen. If eight pounds of oxygen and one pound of hydrogen are mixed and ignited by a flame, they combine with a violent explosion and form nine pounds of limpid water. If this water is now sufficiently heated, *e.g.* by passing it slowly through a red-hot tube, it quietly separates into its original gases—namely, eight pounds of oxygen, one of hydrogen.

Should we deal with these gases by volumes, instead of pounds, then three cubic feet of the combined gases, of which two are hydrogen, and one of oxygen, if mixed and exploded form water; but the water produced only occupies the two thousand six hundredth part of the space filled by the combined gases prior to the explosion.

THE AIR, the atmosphere we breathe, is composed of four parts of nitrogen gas and one part of oxygen. For example, if we mix four gallons of nitrogen and one of oxygen, we have five gallons of air. It is the oxygen that supports life; the nitrogen is simply a diluent or spreader.

New Amendment of the Design Patent Law.

Under the old law, the Supreme Court held that in the case, for example, of a carpet manufacturer who complained of an infringement of his design or pattern of carpet, the complainant must clearly prove what portion of the damage, or what portion of the profit made by the infringer, was due to the use of the patented design. It was practically impossible to make this showing. Hence the infringer could imitate the patented design without liability, and the law was a nullity.

Under the provisions of the new law, the infringer is obliged to pay the sum of $250 in any event; and if his profits are more than that sum, he is compelled, in addition, to pay all excess of profits above $250 to the patentee.

The following is the text of the new law;

Be it enacted by the Senate and House of Representatives of the United States of America in Congress assembled, That hereafter, during the term of letters patent for a design, it shall be unlawful for any person other than the owner of said letters patent, without the license of such owner, to apply the design secured by such letters patent, or any colorable imitation thereof, to any article of manufacture for the purpose of sale, or to sell or expose for sale any article of manufacture to which such design or colorable imitation shall, without the license of the owner, have been applied, knowing that the same has been so applied. Any person violating the provisions, or either of them, of this section shall be liable in the amount of two hundred and fifty dollars; and in case the total profit made by him from the manufacture or sale, as aforesaid, of the article or articles to which the design, or colorable imitation thereof, has been applied, exceeds the sum of two hundred and fifty dollars, he shall be further liable for the excess of such profit over and above the sum of two hundred and fifty dollars; and the full amount of such liability may be recovered by the owner of the letters patent, to his own use, in any circuit court of the United States having jurisdiction of the parties, either by action at law or upon a bill in equity for an injunction to restrain such infringement.

SEC. 2.—That nothing in this act contained shall prevent, lessen, impeach, or avoid any remedy at law or in equity which any owner of letters patent for a design, aggrieved by the infringement of the same, might have had if this act had not been passed; but such owner shall not twice recover the profit made from the infringement.

Approved, February 4, 1887.

INDEX.

	PAGE
Air	52, 143
Appeals	26
Assignments of Patents	27, 59, 66
Attorneys	60
Batteries, Electric	101
Belgian Patents	46
British Patents	45
Canadian Patents	45
Capitol of the U. S	14
Caveats	25
Cement Paint	141
Census of the United States	3
Charcoal, Properties of	119
Circle, Problems	143
Coal, Power of	41
Concrete	141
Copies of Patents	29
Copper	132
Copying Ink	128
Copyright Laws, the	93
Copyrights for Books, etc	33
Copyrights for Labels	32
Crank, Substitute	119
Design Patents	28, 58, 144
Distinguished Inventors	104
Drawings for Patents	54
Electric Batteries, Magnets	101
Electrical Conductors, Table	136
Electricity, Speed of	35
Employers, Employes' rights	35
Engines, Best Coal for	41
English Patents	45
Examination, Official	56
Expansion, Force of	41
Fees, Official, Table of	86
Foreign Patents	44
Forms, various	60
Freezing Mixtures	142
French Patents	46
Friction	130
Geometry, Practical	133
German Patents	46
Gestation	131
Glues, Liquid	142
Going to Washington	23
Gunpowder	129
Heat Conductors	136
Heat, Effects of, Table	136
Heat, Mechanical Equivalent	128
Hints on Sale of Patents	47
Hold the Fort	51
Horse-Power	118
How to Invent	126
How to sell Patents	47
Ice-boats, Velocity of	31
Illumination, Rule for	33
Incubation	131
Infringements	26, 57
Interferences	57
Inventors, Distinguished	103
Knots	120, 121
Labels, Copyrights for	32, 99
Largest and Best	44

	PAGE
License, Form for	67
Light, Velocity, etc	32
Lightning-Rods	141
Mechanical Movements	107
Minerals removed by Crops	142
Model Room, Patent Office	43
Models	22, 56
Molecules	125
Multum in Parvo	112
Occupations in U. S., Tables	137
Patent, Cost of	20
Patent, How much worth	15
Patent Laws, Directions	15
Patent Laws of the U. S	69
Patent Office, View of	24
Patent Office, Washington	43
Patent Sellers, Professional	50
Patented Articles, Stamping	78
Patents, Forms for	60
Patents, Hints on Sale	47
Patents, How to obtain	16
Penalty for Stamping	78
Petition for Patent, Form	60
Plaster of Paris, to Harden	143
Platinum	52
Power of Water	118
Practice, Patent Office, Rules	53
Preliminary Examination	17
Questions and Answers	34
Railways of U. S	14
Recipes, Useful	143
Reissues	26, 57
Rejected Applications	25
Rivers of U. S	14
Royalty	47
Rules of Practice	53
Sale of Patents, Hints	47
Scientific Amer. Offices	36, 38, 40
Scientific Amer. Office, Wash	40
Soap Bubbles	132
Sound	41
Specific gravity	130
State Laws concern'g Pat'nts	52
Steamboats, Small	132
Steam Engine, History	127
Steam Engine, the	135
Steam Navigation of U. S	14
Steam, Pressure of, Table	136
Substitute for the Crank	119
Taxes, none on Patents	40
Telegraph Lines, U. S	14
Trade Marks	30, 87, 90
Underdraining	141
U. S., Length, Breadth	14
U. S.. Map of	18, 19
Useful Recipes, Facts	143
Water, Composition of	143
Water-power, Rule	118
Weights, Measures	122, 123, 124
Weights, Substances, Tables	130
What Security have I?	20
Will it Pay?	40
Wind, Velocity and Force	129

THE SCIENTIFIC AMERICAN
Cyclopedia
of Receipts,
NOTES AND QUERIES.
650 Pages.   Price $5.

This splendid work contains a careful compilation of the most useful Receipts and Replies given in the Notes and Queries of correspondents as published in the SCIENTIFIC AMERICAN during nearly half a century past; together with many valuable and important additions.

Over Twelve Thousand selected receipts are here collected; nearly every branch of the useful arts being represented. It is by far the most comprehensive volume of the kind ever placed before the public.

The work may be regarded as the product of the studies and practical experience of the ablest chemists and workers in all parts of the world; the information given being of the highest value, arranged and condensed in concise form, convenient for ready use.

Almost every inquiry that can be thought of, relating to formulæ used in the various manufacturing industries, will here be found answered.

Instructions for working many different processes in the arts are given. How to make and prepare many different articles and goods is set forth.

Those who are engaged in any branch of industry probably will find in this book much that is of practical value in their respective callings.

Those who are in search of independent business or employment, relating to the home manufacture and sale of useful articles, will find in it hundreds of most excellent suggestions.

www.ingramcontent.com/pod-product-compliance
Lightning Source LLC
Chambersburg PA
CBHW030357170426
43202CB00010B/1398